T0205494

Studies in Computational Intelligence

Volume 720

Series editor

Janusz Kacprzyk, Polish Academy of Sciences, Warsaw, Poland
e-mail: kacprzyk@ibspan.waw.pl

About this Series

The series "Studies in Computational Intelligence" (SCI) publishes new developments and advances in the various areas of computational intelligence—quickly and with a high quality. The intent is to cover the theory, applications, and design methods of computational intelligence, as embedded in the fields of engineering, computer science, physics and life sciences, as well as the methodologies behind them. The series contains monographs, lecture notes and edited volumes in computational intelligence spanning the areas of neural networks, connectionist systems, genetic algorithms, evolutionary computation, artificial intelligence, cellular automata, self-organizing systems, soft computing, fuzzy systems, and hybrid intelligent systems. Of particular value to both the contributors and the readership are the short publication timeframe and the worldwide distribution, which enable both wide and rapid dissemination of research output.

More information about this series at http://www.springer.com/series/7092

Omid Bozorg-Haddad
Editor

Advanced Optimization by Nature-Inspired Algorithms

 Springer

Editor
Omid Bozorg-Haddad
Department of Irrigation & Reclamation
  Engineering
College of Agriculture & Natural Resources,
  University of Tehran
Karaj
Iran

ISSN 1860-949X   ISSN 1860-9503   (electronic)
Studies in Computational Intelligence
ISBN 978-981-13-5345-1   ISBN 978-981-10-5221-7   (eBook)
DOI 10.1007/978-981-10-5221-7

Printed on acid-free paper

This Springer imprint is published by Springer Nature
The registered company is Springer Nature Singapore Pte Ltd.
The registered company address is: 152 Beach Road, #21-01/04 Gateway East, Singapore 189721, Singapore

I like to dedicate this book to my loving parents.

Omid Bozorg-Haddad

Preface

From the early 1990s, the introduction of the term *"Computational Intelligence"* (CI) highlighted the potential applicability of this field. One of the preliminary applications of the field was in the realm of optimization. Undoubtedly, the tasks of design and operation of systems can be approached systematically by the application of optimization. And while in most real-life problems, including engineering problems, application of the classical optimization techniques were limited due to the complex nature of the decision space and numerous variables, and the CI-based optimization techniques, which imitated the nature as a source of inspiration, have proven quite useful. Consequently, during the last passing decades, a considerable number of novel nature-based optimization algorithms have been proposed in the literature. While most of these algorithms hold considerable promise, a majority of them are still in their infancy. For such algorithms to bloom and reach their full potential, they should be implemented in numerous optimization problems, so that not only their most suitable sets of optimization problems are recognized, but also adaptive strategies need to be introduced to make them more suitable for wider sets of optimization problems. For that, this book specifically aimed to introduce some of these potential nature-based algorithms that could be useful for multidisciplinary students including those in aeronautic engineering, mechanical engineering, industrial engineering, electrical and electronic engineering, chemical engineering, civil engineering, computer science, applied mathematics, physics, economy, biology, and social science, and particularly those pursuing postgraduate studies in advanced subjects. Chapter 1 of the book is a review of the basic principles of optimization and nature-based optimization algorithms. Chapters 2–15 are respectively dedicated to Cat Swarm Optimization (CSO), League Championship Algorithm (LCA), Anarchies Society Optimization (ASO), Cuckoo Optimization Algorithm (COA), Teacher-Learning-Based Optimization (TLBO), Flower Pollination Algorithm (FPA), Krill Herd Algorithm (KHA), Grey Wolf Optimization (GWO), Shark Smell Optimization (SSO), Ant Lion Optimization (ALO), Gradient Evolution (GE), Moth-Flame Optimization (MFO), Crow Search Algorithm (CSA), and Dragonfly Algorithm (DA). The order of the chapters corresponds to the order of chronological appearance of these algorithms, from earlier algorithms to newly introduced ones.

Each chapter describes a specific algorithm and starts with a brief literature review of its development and subsequent modification since the time of inception. This is followed by the presentation of the basic concept on which the algorithm is based and the steps of the algorithm. Each chapter closes with a pseudocode of the algorithm.

Karaj, Iran   Omid Bozorg-Haddad

Contents

About the Editor

Omid Bozorg-Haddad is Professor in the Department of Irrigation and Reclamation Engineering at the University of Tehran, Iran. His teaching and research interests include water resources and environmental systems analysis, planning, and management as well as application of optimization algorithms in water related systems. He has published more than 100 articles in peer-reviewed journals and 100 papers in conference proceedings. He has also supervised more than 50 M.Sc. and Ph.D. students.

List of Figures

List of Tables

Chapter 1
Introduction

Babak Zolghadr-Asli⊙, Omid Bozorg-Haddad and Xuefeng Chu

Abstract In this chapter, some general knowledge relative to the realm of nature-inspired optimization algorithms (NIOA) is introduced. The desirable merits of these intelligent algorithms and their initial successes in many fields have inspired researchers to continuously develop such revolutionary algorithms and implement them to solve various real-world problems. Such a truly interdisciplinary environment of the research and development provides rewarding opportunities for scientific breakthrough and technology innovation. After a brief introduction to computational intelligence and its application in optimization problems, the history of the NIOA was reviewed. The relevant algorithms were then categorized in different manners. Finally, one the most groundbreaking theorems regarding the nature-inspired optimization techniques was briefly discussed.

1.1 Introduction

Artificial intelligence (AI) refers to any sort of intelligence that is exhibited by machines. The term of computational intelligence (CI), a branch of AI, was coined by Bezdek in the early 1990s (Bezdek 1992), which inspired the development of a

B. Zolghadr-Asli · O. Bozorg-Haddad (✉)
Department of Irrigation and Reclamation Engineering, Faculty of Agricultural Engineering and Technology, College of Agriculture and Natural Resources, University of Tehran, 3158777871 Karaj, Iran
e-mail: obhaddad@ut.ac.ir

B. Zolghadr-Asli
e-mail: zolghadrbabak@ut.ac.ir

X. Chu
Department of Civil and Environmental Engineering, North Dakota State University, Dept 2470, Fargo, ND 58108-6050, USA
e-mail: xuefeng.chu@ndsu.edu

© Springer Nature Singapore Pte Ltd. 2018   1
O. Bozorg-Haddad (ed.), *Advanced Optimization by Nature-Inspired Algorithms*, Studies in Computational Intelligence 720, DOI 10.1007/978-981-10-5221-7_1

new field dedicated to computer-based intelligence. In principle, CI consists of any science-supported approaches and technologies for analyzing, creating, and developing intelligent systems (Xing and Gao 2014). Unlike AI that depends upon the knowledge derived from human expertise, CI, however, mostly relies on the collection of numerical data and a set of nature-inspired computational paradigms (Du and Swamy 2016). As evidenced by numerous studies in various fields, CI was able to flourish and achieve an unprecedented popularity during the past several decades. While, once the CI's primary focus was on artificial neural network (ANN), fuzzy logic (FL), multi-agent system (MAS), and few optimization algorithms including: evolutionary algorithms (EA) [e.g., genetic algorithm (GA), genetic programming (GP), evolutionary programming (EP), and evolutionary strategy (ES)], artificial immune systems (AIS), simulated annealing (SA), Tabu search (TS), as well as two variants of swarm intelligence (SI) [i.e., ant colony optimization (ACO) and particle swarm optimization (PSO)] (Xing and Gao 2014), nowadays CI has expanded to a vast domain that covers almost every science and engineering field.

One of the major subjects of CI is, as previously mentioned, the study of unorthodox optimization techniques. It is no exaggeration to say that optimization has surrounded our everyday life, from complex engineering problems to business or even the holiday planning. What bonds these activities together is that essentially they aim to utilize a finite commodity (say, a limited resource, finance, or even time) to achieve certain goals or objectives. Mathematically speaking, optimization searches the decision space in pursuit of an array of decision variables that could produce the maximum acceptability considering the initial goal (objective function). Since its introduction, CI has been able to make a considerable contribution to the solving techniques of such optimization problems due to the development of various efficient search algorithms.

Traditionally, optimization problems were solved either using calculus-based methods, random-based search, or in some cases enumerative searching techniques (Du and Swamy 2016). Broadly speaking, such optimization techniques can generally be divided into derivative and nonderivative methods, depending on whether or not the derivatives of the objective function(s) are required in the optimization process. Derivative methods are calculus-based methods, which are based on a gradient search (also known as the steepest search method). The Newton's method, Gauss–Newton method, quasi-Newton methods, the trust-region method, and Levenberg–Marquardt method are a few examples of such techniques. These classical methods used to dominate the field of optimization. While efficient in solving a wide range of optimization problems, when the number of the decision variables grow large enough, or even when the decision space is discreet, which is the case in most if not all practical optimization problems, these techniques are unable to illustrate acceptable performances. Indeed, the aforementioned describes one of the major advantages of CI-based optimization techniques over traditional ones.

1.2 Optimization: Core Principles and Technical Terms

Assume that a group of amateur climbers decide to summit the highest mountain in a previously unknown and hilly territory. Indeed, searching for the summit is not unlike searching for the optimal solution, for, in fact, the landscape represents the decision space, while the highest mountain embodies the global optimum. But how does one even begin to search such a vast area? One initial answer would be to map out the entire landscape. However, this would be a both time- and energy-consuming task. Perhaps, an alternative would be to search the area in a random-based manner. While it would definitely help to save both time and energy, ultimately as one can expect, it would not be an efficient strategy as well, for there would be no guarantee to reach the highest mountain. A more intelligent alternative would be to directly climb up the steepest cliff, assuming that the highest mountain is more likely to have a steeper slope. Essentially, this strategy represents the core principle of classical optimization techniques. While efficient in many cases, if the climbers' path would be interrupted by cliffs (discrete decision space), for instance, this strategy would not be efficient to locate the highest mountain. Additionally, in a hilly landscape, climbers could deceitfully climb to the top of a mountain which in essential stands above the neighboring mountains, while in fact, it is not the highest mountain in the entire area. This problem is known in technical terms as trapping in local optima.

Alternatively, the climbers could do a random walk in the area, while looking for some clues. Such hybrid strategies are formed using a combination of random-based searching and an adaptive strategy, which is usually inspired by nature. In fact, that is the description of CI-based optimization algorithms. Subsequently, such searches could be conducted while the group maintains to stick with one another (perform as an individual climber). The group members can also spread out and share their gained information with each other, while the searching proceeds further on. Technically speaking, the former strategy is better known as single-point optimization technique, while the latter strategy represents population-based optimization algorithms. The single-point strategies are also known as trajectory optimization algorithms, for the optimization process would provide a path that could lead to the optimum point, in this case, the highest mountain in the area.

Additionally, the climbers, either as one group or separate individuals, could investigate the area only using the information currently at hand. An alternative would be to take a record and map out some previously encountered locations. In a technical term, the second strategy is called memory using algorithms. While such a strategy is more efficient in most cases, if the population of the climbers increases or the landscape is vast enough, storing such massive information could potentially turn into a major problem.

Ultimately, the core principle of all CI-based optimization algorithms, which are better known as metaheuristic algorithms, is a way of trial and error to produce an acceptable solution to a complex problem in a reasonably practical time (Yang 2010). While the complexity of the practical optimization problem is in favor of implementing such optimization algorithms, there could also be no guarantee that the best solution (global optimum) can be spotted using such techniques.

1.3 Brief History of CI-Based Optimization Algorithms

Despite their novel and ubiquitous nature, implementation of the CI-based optimization algorithms is indeed a relatively new technique, though it is difficult to pinpoint when the whole story began. Accordingly, Allen Turing was perhaps the first person to implement the CI-based optimization algorithms (Yang 2010). Evidently, during World War II, while trying to break the German-designed encrypting machine called Enigma, Turing developed an optimization technique which he later named heuristic search as it could be expected to work most of the times, while there was no actual guarantee for a successful performance in each trial. Turing was later recruited to the national physical laboratory, UK, where he set out his design for the automatic computing engine. Later on, he outlined his innovative idea of machine intelligence and learning neural networks and evolutionary algorithms in an NPL report on intelligent machinery (Yang 2010).

The CI-based optimization techniques bloomed during the 1960s and 1970s. In the early 1960s, John Holland and his collaborators at the University of Michigan developed the genetic algorithms (GA) (Goldberg and Holland 1988). In essence, a GA is a search method based on the abstraction of Darwin's theory of evolution and natural selection of biological systems, which are represented in the mathematical operators. Holland's preliminary studies were showing promising results, while he continued to further develop his technique by introducing novel and efficient agents to his algorithms which were named crossover and mutation, although his seminal book summarizing the development of the genetic algorithm was not published until 1975 (Yang 2010). Holland's work inspired many to further develop and adopt similar methods in their research, which benefited from a similar basic principle in numerous and colorful fields. For instance, while Holland and his collogues were trying to develop their revolutionary method, Ingo Rechenberg and Hans-Paul Schwefel at the Technical University of Berlin introduced another novel optimization technique for solving aerospace engineering problems, which they later named evolutionary strategy (Back et al. 1997). In 1966, (Fogel et al. 1966) developed an evolutionary programming technique by representing the solution as finite state machines and randomly mutating one of these machines. The above innovative idea and method have evolved into a wider area that became known as evolutionary algorithms (EAs) (Yang 2010).

In the early 1990s and in another great leap forward in the field of CI-based optimization algorithms, Marco Dorigo finished his Ph.D. thesis on optimization and nature-inspired algorithms, in which he described his innovative work on ant colony optimization (ACO) (Dorigo and Blum 2005). This search technique was inspired by the swarm intelligence of social ants using the pheromone to trace the food sources. Slightly later, in 1995, the particle swarm optimization (PSO) was proposed by an

American social physiologist James Kennedy, and engineered by Russell C. Eberhart, which shall be considered as another significant progress in the field of nature-inspired optimization techniques (Poli et al. 2007). In essence, PSO is an optimization algorithm inspired by the swarm intelligence of the school of fish or birds. Multiple agents or particles swarm around the search space starting from an initial random guess. The swarm members communicate to one another by sharing the current best answer, which in turn, enables the algorithm to find the optimum solutions. Since then, many attempts have been made to mimic and imitate the natural beings in the searching process of optimization techniques. During the past decades, numerous novel nature-inspired optimization algorithms have been proposed to advance the CI-based optimization techniques (Xing and Gao 2014).

1.4 Classification of CI-Based Optimization Algorithms

Perhaps, it could be beneficial for studding reasons to categorize and classify the CI-based optimization techniques, although depending on specific viewpoints many classifications are possible. While some of these classifications are quite technical and in some cases, even based upon vague characteristics, more general classifications could help better understand the core principles behind such algorithms.

Intuitively, one can categorize the CI-based algorithms using the number of searching agents. This characteristic would divide the algorithms into two major categories: (1) population-based algorithms, and (2) single-point algorithms. The algorithms working with a single agent are called trajectory methods. The population-based algorithms, on the other hand, perform search processes using several agents simultaneously.

As stated in previous sections, some CI-based algorithms can keep a record of previously inspected arrays of decision variable in the search space. Such algorithms are known as the memory using algorithms. On the contrary, the algorithms, known as memory-less algorithms, do not memorize the previously encountered locations in the search space. The methods of making use of memory in a CI-based optimization algorithm can be further divided into short term and long-term memory using algorithms. The former usually keeps track of recently performed moves, visited solutions or, in general, decisions taken, while the latter is usually an accumulation of synthetic parameters about the search. The use of memory is nowadays recognized as one of the fundamental elements of a powerful CI-based optimization algorithm (Blum and Roli 2003).

As mentioned earlier on, one of the challenges in complex optimization problems is to avoid trapping in local optima, which has been a major problem for classical optimization techniques. To overcome this problem, some CI-based optimization algorithms would modify their objective function during the optimization process. Such optimization techniques are known as dynamic algorithms. The alternative would be to keep the objective function as is during the optimization process. This is the characteristic of static algorithms.

Finally, a more general, yet sometimes vague, classification of the CI-based optimization algorithms is based on their sources of inspiration. Based on their origins, the CI-based optimization algorithms could be classified as nature-based and non-nature-based optimization algorithms. While such a classification could be confusing, for there are cases in which it is rather difficult to clearly attribute an algorithm to one of the two classes, yet, such classification is the most widely accepted classification for the CI-based algorithms. The subject of this book is nature-based optimization algorithms. Generally speaking, any biological-, physical-, or even chemical-based algorithm that somehow imitates the nature can be categorized as a nature-based optimization algorithm. Also, it should be mentioned that such a categorizing method for the CI-based optimization algorithms has no contradiction with any previously introduced categories. For instance, a nature-based optimization algorithm could also be a population-based or single-point optimization algorithm.

It is advantageous to remember that two major components of any nature-based algorithms are intensification and diversification. Diversification means to generate diverse solutions so as to explore the search space, while intensification means to focus on the search in a local region by exploiting the area surrounding the current good solutions. While in most cases through a random-based component (diversification), nature-based optimization algorithms attempt to thoroughly investigate the decision space and divert the solutions to be trapped in local optima, the intensification component ensures the convergence of the algorithm to what is more likely to be the global optimum solution. The good combination of these two major components will usually ensure that the global optimality is achievable (Yang 2010).

1.5   No Free Lunch Theorem: The Reason Behind New Algorithms

In the late 1990s, (Wolpert and Macready 1997) proposed a controversial, groundbreaking no free lunch theorem for optimization. Researchers intuitively believed that there exist some universally beneficial robust algorithms for optimization. However, the no free lunch theorem clearly states that if an algorithm outperforms others for some optimization functions, there will be other optimization functions in which the other algorithms shall be better than the aforementioned algorithm. In other words, if all possible function spaces are to be considered, on average, all CI-based algorithms, and subsequently all nature-based algorithms should perform equally well. And thus, ultimately, there could be no universally better algorithm.

While no free lunch theorem clearly dismisses the myth of an ultimate optimization algorithm, there is a silver lining sub-layer to this theorem, as well. Though all algorithms are to perform equally on the average over all possible function space, yet as evidenced by numerous studies, such algorithms may and will outperform one another in specific sets of optimization problems. In addition, the algorithm developments are now focused on finding the best and efficient algorithm for a specific set of optimization problems. Ultimately, instead of aiming to design a perfect solver for all the problems, algorithms are developed to solve most types of problems. As a result, during the last several decades, a considerable number of novel nature-based optimization algorithms have been proposed. While most of the algorithms hold considerable promise, a majority of them are still in their infancy. For such algorithms to bloom and reach their full potential, they should be applied to numerous optimization problems, so that not only they are tested for their most suitable sets of optimization problems, but also adaptive strategies are introduced to make them more suitable for wider sets of optimization problems. For that, this book is specifically aimed to introduce some of the potential nature-based algorithms that can be useful in multidisciplinary studies including those in aeronautic engineering, mechanical engineering, industrial engineering, electrical and electronic engineering, chemical engineering, civil and environmental engineering, computer science and engineering, applied mathematics, physics, economy, biology, and social science, and particularly the postgraduate studies in advanced subjects.

1.6   Conclusion

In this chapter, a brief background of CI, which consists of any science-supported approaches and technologies for analyzing, creating, and developing intelligent systems has been discussed from an introductory perspective. A major subject of CI is to develop nature-based optimization techniques. For decades the nature-based optimization methods have proved their ability to solve complex practical optimization problems. Nowadays there are many novel nature-based optimization techniques. However, according to the no free lunch theorem such algorithms do not have particular advantages over one another, but rather could outperform each other for particular sets of optimization problems. As a result, this book aims to introduce some of the potentially useful algorithms so that the right audience could check their authentic outperformances in applications and perhaps modify or select the best algorithm for any particular optimization problem in hand.

References

Back, T., Hammel, U., & Schwefel, H. P. (1997). Evolutionary computation: Comments on the history and current state. *IEEE Transactions on Evolutionary Computation, 1*(1), 3–17.
Bezdek, J. C. (1992). On the relationship between neural networks, pattern recognition and intelligence. *International Journal of Approximate Reasoning, 6*(2), 85–107.

Blum, C., & Roli, A. (2003). Metaheuristics in combinatorial optimization: Overview and conceptual comparison. *ACM Computing Surveys (CSUR), 35*(3), 268–308.

Dorigo, M., & Blum, C. (2005). Ant colony optimization theory: A survey. *Theoretical Computer Science, 344*(2–3), 243–278.

Du, K. L., & Swamy, M. N. S. (2016). *Search and optimization by metaheuristics*. Switzerland: Springer Publication.

Fogel, L. J., Owens, A. J., & Walsh, M. J. (1966). Intelligent decision making through a simulation of evolution. *Behavioral Science, 11*(4), 253–272.

Goldberg, D. E., & Holland, J. H. (1988). Genetic algorithms and machine learning. *Machine Learning, 3*(2), 95–99.

Poli, R., Kennedy, J., & Blackwell, T. (2007). Particle swarm optimization. *Swarm Intelligence, 1*(1), 33–57.

Wolpert, D. H., & Macready, W. G. (1997). No free lunch theorems for optimization. *IEEE Transactions on Evolutionary Computation, 1*(1), 67–82.

Xing, B., & Gao, W. J. (2014). *Innovative computational intelligence: A rough guide to 134 clever algorithms*. Cham, Switzerland: Springer Publication.

Yang, X. S. (2010). *Nature-inspired metaheuristic algorithms*. Frome, UK: Luniver Press.

Chapter 2
Cat Swarm Optimization (CSO) Algorithm

Mahdi Bahrami, Omid Bozorg-Haddad and Xuefeng Chu

Abstract In this chapter, a brief literature review of the Cat Swarm Optimization (CSO) algorithm is presented. Then the natural process, the basic CSO algorithm iteration procedure, and the computational steps of the algorithm are detailed. Finally, a pseudo code of CSO algorithm is also presented to demonstrate the implementation of this optimization technique.

2.1 Introduction

Optimization algorithms based on the Swarm Intelligence (SI) were developed for simulating the intelligent behavior of animals. In these modeling systems, a population of organisms such as ants, bees, birds, and fish are interacting with one another and with their environment through sharing information, resulting in use of their environment and resources. One of the more recent SI-based optimization algorithms is the Cat Swarm Optimization (CSO) algorithm which is based on the behavior of cats. Developed by Chu and Tsai (2007), the CSO algorithm and its varieties have been implemented for different optimization problems. Different variations of the algorithm have been developed by researchers. Tsai et al. (2008)

M. Bahrami · O. Bozorg-Haddad (✉)
Department of Irrigation and Reclamation Engineering, Faculty of Agricultural Engineering and Technology, College of Agriculture and Natural Resources, University of Tehran, Karaj, Tehran 31587-77871, Iran
e-mail: obhaddad@ut.ac.ir

M. Bahrami
e-mail: m.bahrami9264@ut.ac.ir

X. Chu
Department of Civil and Environmental Engineering, North Dakota State University, Dept 2470, Fargo, ND 58108-6050, USA
e-mail: xuefeng.chu@ndsu.edu

© Springer Nature Singapore Pte Ltd. 2018
O. Bozorg-Haddad (ed.), *Advanced Optimization by Nature-Inspired Algorithms*,
Studies in Computational Intelligence 720, DOI 10.1007/978-981-10-5221-7_2

9

presented a parallel structure of the algorithm (i.e., parallel CSO or PCSO). They further developed an enhanced version of their PCSO (EPCSO) by incorporating the Taguchi method into the tracing mode process of the algorithm (Tsai et al. 2012). The binary version of CSO (BCSO) was developed by Sharafi et al. (2013) and applied to a number of benchmark optimization problems and the zero–one knapsack problem. The chaotic cat swarm algorithm (CCSA) was developed by Yang et al. (2013a). Using different chaotic maps, the seeking mode step of the algorithm was improved. Based on the concept of homotopy, Yang et al. (2013b), proposed the homotopy-inspired cat swarm algorithm (HCSA) in order to improve the search efficiency. Lin et al. (2014a) proposed a method to improve CSO and presented the Harmonious-CSO (HCSO). Lin et al. (2014b) introduced a modified CSO (MCSO) algorithm capable of improving the search efficiency within the problem space. The basic CSO algorithm was also integrated with a local search procedure as well as the feature selection of support vector machines (SVMs). This method changed the concept of cat alert surroundings in the seeking mode of CSO algorithm. By dynamically adjusting the mixture ratio (MR) parameter of the CSO algorithm, Wang (2015) enhanced CSO algorithm with an adaptive parameter control. A hybrid cat swarm optimization method was developed by Ojha and Naidu (2015) through adding the invasive weed optimization (IWO) algorithm to the tracing mode of the CSO algorithm.

Several other authors have used CSO algorithm in different fields of research on optimization problems. Lin and Chien (2009) constructed the CSO algorithm + SVM model for data classification through integrating cat swarm optimization into the SVM classifier. Pradhan and Panda (2012) proposed a new multiobjective evolutionary algorithm (MOEA) by extending CSO algorithm. The MOEA identified the non-dominated solutions along the search process using the concept of Pareto dominance and used an external archive for storing them. Xu and Hu (2012) presented a CSO-based method for a resource-constrained project scheduling problem (RCPSP). Saha et al. (2013) applied CSO algorithm to determine the optimal impulse response coefficients of FIR low pass, high pass, bandpass, and band stop filters to meet the respective ideal frequency response characteristics. So and Jenkins (2013) used CSO for Infinite Impulse Response (IIR) system identification on a few benchmarked IIR plants. Kumar et al. (2014) optimized the placement and sizing of multiple distributed generators using CSO. Mohamadeen et al. (2014) compared the binary CSO with the binary PSO in selecting the best transformer tests that were utilized to classify transformer health, and thus to improve the reliability of identifying the transformer condition within the power system. Guo et al. (2015) proposed an improved cat swarm optimization algorithm and redefined some basic CSO concepts and operations according to the assembly sequence planning (ASP) characteristics. Bilgaiyan et al. (2015) used the cat swarm-based multi-objective optimization approach to schedule workflows in a cloud computing environment which showed better performance, compared with the multi-objective particle swarm optimization (MOPSO) technique. Amara et al. (2015) solved the problem of wind power system design reliability optimization using CSO, under the performance and cost constraints. Meziane et al. (2015) optimized the electric power distribution of a solar system by determining the optimal topology among various alternatives using CSO. The results showed a better performance than the binary CSO. Ram et al. (2015) studied a 9-ring time-modulated concentric circular antenna array (TMCCAA) with isotropic elements based on CSO, for reduction of side lobe level and improvement in the directivity. Crawford et al. (2016) solved a bi-objective set covering problem using the binary cat swarm optimization algorithm. In order to achieve higher overall system reliability for a large-scale primary distribution network, Majumder and Eldho (2016) examined the effectiveness of CSO for groundwater management problems, by coupling it with the analytic element method (AEM) and the reverse particle tracking (RPT) approach. The AEM-CSO model was applied to a hypothetical unconfined aquifer considering two different objectives: maximization of the total pumping of groundwater from the aquifer and minimization of the total pumping costs. Mohapatra et al. (2016) used kernel ridge regression and a modified CSO-based gene selection system for classification of microarray medical datasets.

2.2   Natural Process of the Cat Swarm Optimization Algorithm

Despite spending most of their time in resting, cats have high alertness and curiosity about their surroundings and moving objects in their environment. This behavior helps cats in finding preys and hunting them down. Compared to the time dedicated to their resting, they spend too little time on chasing preys to conserve their energy. Inspired by this hunting pattern, Chu and Tsai (2007) developed CSO with two modes: "seeking mode" for when cats are resting and "tracing mode" for when they are chasing their prey. In CSO, a population of cats are created and randomly distributed in the M-dimensional solution space, with each cat representing a solution. This population is divided into two subgroups. The cats in the first subgroup are resting and keeping an eye on their surroundings (i.e., seeking mode), while the cats in the second subgroup start moving around and chasing their preys (i.e., tracing mode). The mixture of these two modes helps CSO to move toward the global solution in the M-dimensional solution space. Since the cats spend too little time in the tracing mode, the number of the cats in the tracing subgroup should be small. This number is defined by using the mixture ratio (MR) which has a small value. After sorting the cats into these two modes, new positions and fitness functions will be available, from which the cat with the best solution will be saved in the memory. These steps are repeated until the stopping criteria are satisfied.

Table 2.1 Characteristics of the CSO algorithm

General algorithm	Cat swarm optimization
Decision variable	Cat's position in each dimension
Solution	Cat's position
Old solution	Old position of cat
New solution	New position of cat
Best solution	Any cat with the best fitness
Fitness function	Distance between cat and prey
Initial solution	Random positions of cats
Selection	–
Process of generating new solution	Seeking and tracing a prey

Following Chu and Tsai (2007), the computational procedures of CSO can be described as follows:

Step 1: Create the initial population of cats and disperse them into the M-dimensional solution space $(X_{i,d})$ and randomly assign each cat a velocity in range of the maximum velocity value $(v_{i,d})$.

Step 2: According to the value of MR, assign each cat a flag to sort them into the seeking or tracing mode process.

Step 3: Evaluate the fitness value of each cat and save the cat with the best fitness function. The position of the best cat (X_{best}) represents the best solution so far.

Step 4: Based on their flags, apply the cats into the seeking or tracing mode process as described below.

Step 5: If the termination criteria are satisfied, terminate the process. Otherwise repeat steps 2 through 5.

Table 2.1 lists the characteristics of the CSO and Fig. 2.1 illustrates the detailed computational steps of the CSO algorithm.

2.2.1 Seeking Mode (Resting)

During this mode the cat is resting while keeping an eye on its environment. In case of sensing a prey or danger, the cat decides its next move. If the cat decides to move, it does that slowly and cautiously. Just like while resting, in the seeking mode the cat observes into the M-dimensional solution space in order to decide its next move. In this situation, the cat is aware of its own situation, its environment, and the choices it can make for its movement. These are represented in the CSO algorithm by using four parameters: seeking memory pool (SMP), seeking range of

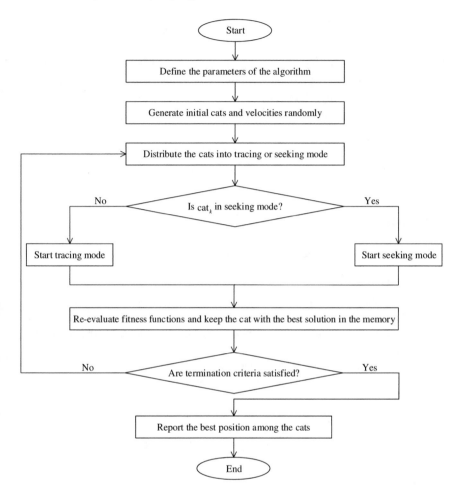

Fig. 2.1 Flowchart of the CSO algorithm

the selected dimension (SRD), counts of dimension to change (CDC), and self-position consideration (SPC) (Chu and Tsai 2007). SMP is the number of the copies made of each cat in the seeking process. SRD is the maximum difference between the new and old values in the dimension selected for mutation. CDC tells how many dimensions will be mutated. All these parameters define the seeking process of the algorithm. SPC is the Boolean variable which indicates the current position of the cat as a candidate position for movement. SPC cannot affect the value of SMP.

Following Chu and Tsai (2007), the process of the seeking mode is described below.

Step 1: Make SMP copies of each cat$_i$. If the value of SPC is true, SMP-1 copies are made and the current position of the cat remains as one of the copies.

Step 2: For each copy, according to CDC calculate a new position by using Eq. (2.1) (Majumder and Eldho 2016)

$$X_{cn} = (1 \pm \text{SRD} \times R) \times X_c \qquad (2.1)$$

in which

X_c current position;
X_{cn} new position; and
R a random number, which varies between 0 and 1.

Step 3: Compute the fitness values (FS) for new positions. If all FS values are exactly equal, set the selecting probability to 1 for all candidate points. Otherwise calculate the selecting probability of each candidate point by using Eq. (2.2).

Step 4: Using the roulette wheel, randomly pick the point to move to from the candidate points, and replace the position of cat$_i$.

$$P_i = \frac{|FS_i - FS_b|}{|FS_{\max} - FS_{\min}|}, \quad \text{where } 0 < i < j \qquad (2.2)$$

where

P_i probability of current candidate cat$_i$;
FS_i fitness value of the cat$_i$;
$FS_{\max}$ maximum value of fitness function;
$FS_{\min}$ minimum value of fitness function; and
$FS_b = FS_{\max}$ for minimization problems and
$FS_b = FS_{\min}$ for maximization problems.

2.2.2   Tracing Mode (Movement)

The tracing mode simulates the cat chasing a prey. After finding a prey while resting (seeking mode), the cat decides its movement speed and direction based on the prey's position and speed. In CSO, the velocity of cat k in dimension d is given by

$$v_{k,d} = v_{k,d} + r_1 \times c_1 (X_{\text{best},d} - X_{k,d}) \qquad (2.3)$$

in which, $v_{k,d}$ = velocity of cat k in dimension d; $X_{\text{best},d}$ = position of the cat with the best solution; $X_{k,d}$ = position of the cat$_k$; c_1 = a constant; and r_1 = a random value in the range of [0,1]. Using this velocity, the cat moves in the M-dimensional decision space and reports every new position it takes. If the velocity of the cat is greater than the maximum velocity, its velocity is set to the maximum velocity. The new position of each cat is calculated by

$$X_{k,d,\text{new}} = X_{k,d,\text{old}} + v_{k,d} \qquad (2.4)$$

in which

$X_{k,d,\text{new}}$ new position of cat k in dimension d; and
$X_{k,d,\text{old}}$ current position of cat k in dimension d.

2.3 Termination Criteria

The termination criterion determines when the algorithm is terminated. Selecting a good termination criterion has an important role to ensure a correct convergence of the algorithm. The number of iterations, the amount of improvement, and the running time are common termination criteria for the CSO.

2.4 Performance of the CSO Algorithm

Chu and Tsai (2007) used six test functions to evaluate the CSO performance and compared the results with the particle swarm optimization (PSO) algorithm and the PSO with weighting factor (PSO-WF). According to the results CSO outperformed PSO and PSO-WF in finding the global best solutions.

2.5   Pseudo Code of the CSO Algorithm

Begin

Input parameters of the algorithm and the initial data

Initialize the cat population X_i ($i = 1, 2, \ldots, n$), v, and SPC

While (the stop criterion is not satisfied or $I < I_{max}$)

Calculate the fitness function values for all cats and sort them

$X_g =$ cat with the best solution

For $i = 1: N$

If $SPC = 1$

Start seeking mode

Else

Start tracing mode

End if

End for i

End while

Post-processing the results and visualization

End

2.6   Conclusion

This chapter described cat swarm optimization (CSO) which is a new swarm-based algorithm. CSO consists of two modes, seeking mode and tracing mode which simulate the resting and hunting behaviors of cats. Each cat has a position in the M-dimensional solution space. The cats' movement toward the optimum solution is based on a flag that sorts them into the seeking or tracing mode, the first one being a slow movement around their environment and the latter being a fast movement toward the global best.

A literature review of CSO was presented, showing the success of the algorithm for different optimization problems, along with different variations of the code developed by other researchers. The flowchart of the CSO along with the pseudo code was also presented in order to make different parts of the algorithm easier to understand. These sources are a good reference point for further exploration of the CSO algorithm.

References

Amara, M., Bouanane, A., Meziane, R., & Zeblah, A. (2015). *Hybrid wind gas reliability optimization using cat swarm approach under performance and cost constraints.* 3rd International Renewable and Sustainable Energy Conference (IRSEC), Marrakech and Ouarzazate, Morocco, 10–13 December.

Bilgaiyan, S., Sagnika, S., & Das, M. (2015). A multi-objective cat swarm optimization algorithm for workflow scheduling in cloud computing environment. Intelligent Computing, Communication and Devices (pp. 73–84). New Delhi, India: Springer.

Chu, S. C., & Tsai, P. W. (2007). Computational intelligence based on the behavior of cats. *International Journal of Innovative Computing, Information and Control, 3*(1), 163–173.

Crawford, B., Soto, R., Caballero, H., Olguín, E., & Misra, S. (2016). *Solving biobjective set covering problem using binary cat swarm optimization algorithm.* The 16th International Conference on Computational Science and Its Applications, Beijing, China, 4–7 July.

Guo, J., Sun, Z., Tang, H., Yin, L., & Zhang, Z. (2015). Improved cat swarm optimization algorithm for assembly sequence planning. *Open Automation and Control Systems Journal, 7,* 792–799.

Kumar, D., Samantaray, S. R., Kamwa, I., & Sahoo, N. C. (2014). Reliability-constrained based optimal placement and sizing of multiple distributed generators in power distribution network using cat swarm optimization. *Electric Power Components and Systems, 42*(2), 149–164.

Lin, K. C., & Chien, H. Y. (2009). *CSO-based feature selection and parameter optimization for support vector machine.* Joint Conferences on Pervasive Computing (JCPC), Taipei, Taiwan, 3–5 December.

Lin, K. C., Zhang, K. Y., & Hung, J. C. (2014a). *Feature selection of support vector machine based on harmonious cat swarm optimization.* Ubi-Media Computing and Workshops (UMEDIA), Ulaanbaatar, Mongolia, 12–14 July.

Lin, K. C., Huang, Y. H., Hung, J. C., & Lin, Y. T. (2014b). Modified cat swarm optimization algorithm for feature selection of support vector machines. *Frontier and Innovation in Future Computing and Communications*, 329–336.

Majumder, P., & Eldho, T. I. (2016). A new groundwater management model by coupling analytic element method and reverse particle tracking with cat swarm optimization. *Water Resources Management, 30*(6), 1953–1972.

Meziane, R., Boufala, S., Amara, M., & Hamzi, A. (2015). *Cat swarm algorithm constructive method for hybrid solar gas power system reconfiguration.* 3rd International Renewable and Sustainable Energy Conference (IRSEC), Marrakech and Ouarzazate, Morocco, 10–13 December.

Mohamadeen, K. I., Sharkawy, R. M., & Salama, M. M. (2014). *Binary cat swarm optimization versus binary particle swarm optimization for transformer health index determination.* 2nd International Conference on Engineering and Technology, Cairo, Egypt, 19–20 April.

Mohapatra, P., Chakravarty, S., & Dash, P. K. (2016). Microarray medical data classification using kernel ridge regression and modified cat swarm optimization based gene selection system. *Swarm and Evolutionary Computation, 28,* 144–160.

Naidu, Y. R., & Ojha, A. K. (2015). A hybrid version of invasive weed optimization with quadratic approximation. *Soft Computing, 19*(12), 3581–3598.

Pradhan, P. M., & Panda, G. (2012). Solving multiobjective problems using cat swarm optimization. *Expert Systems with Applications, 39*(3), 2956–2964.

Ram, G., Mandal, D., Kar, R., & Ghoshal, S. P. (2015). Cat swarm optimization as applied to time-modulated concentric circular antenna array: Analysis and comparison with other stochastic optimization methods. IEEE Transactions on Antennas and Propagation, 63(9), 4180–4183.

Saha, S. K., Ghoshal, S. P., Kar, R., & Mandal, D. (2013). Cat swarm optimization algorithm for optimal linear phase FIR filter design. *ISA Transactions, 52*(6), 781–794.

Sharafi, Y., Khanesar, M. A., & Teshnehlab, M. (2013). *Discrete binary cat swarm optimization algorithm.* In Computer, Control & Communication (IC4). 3rd IEEE International Conference on Computer, Control & Communication (IC4), Karachi, Pakistan, 25–26 September.

So, J., & Jenkins, W. K. (2013). *Comparison of cat swarm optimization with particle swarm optimization for IIR system identification.* Asilomar Conference on Signals, Systems and Computers, Pacific Grove, CA, USA, 6–9 November.

Tsai, P. W., Pan, J. S., Chen, S. M., Liao, B. Y., & Hao, S. P. (2008). *Parallel cat swarm optimization.* International Conference on Machine Learning and Cybernetics, Kunming, China, 12–15 July.

Tsai, P. W., Pan, J. S., Chen, S. M., & Liao, B. Y. (2012). Enhanced parallel cat swarm optimization based on the Taguchi method. *Expert Systems with Applications, 39*(7), 6309–6319.

Wang, J. (2015). A new cat swarm optimization with adaptive parameter control. *Genetic and Evolutionary Computing*, 69–78.

Xu, L., & Hu, W. B. (2012). Cat swarm optimization-based schemes for resource-constrained project scheduling. *Applied Mechanics and Materials, 220,* 251–258.

Yang, S. D., Yi, Y. L., & Shan, Z. Y. (2013a). Chaotic cat swarm algorithms for global numerical optimization. *Advanced Materials Research, 602,* 1782–1786.

Yang, S. D., Yi, Y. L., & Lu, Y. P. (2013b). Homotopy-inspired cat swarm algorithm for global optimization. *Advanced Materials Research, 602,* 1793–1797.

Chapter 3
League Championship Algorithm (LCA)

Hossein Rezaei, Omid Bozorg-Haddad and Xuefeng Chu

Abstract This chapter briefly describes the league championship algorithm (LCA) as one of the new evolutionary algorithms. In this chapter, a brief literature review of LCA is first presented; and then the procedure of holding a common league in sports and its rules are described. Finally, a pseudo code of LCA is presented.

3.1 Introduction

The league championship algorithm (LCA) is one of the new evolutionary algorithms (EA) for finding global optimum in a continuous search space first proposed by Kashan (2009, 2011) developed the basic LCA for solving a constrained optimization benchmark problem. The results demonstrated that LCA is a very competitive algorithm for constrained optimization problems. Kashan et al. (2012) modified the basic LCA by using two halves analysis, instead of the post-match SWOT analysis. The performance of the more realistic modified LCA (RLCA) was compared with those of the particle swarm optimization (PSO) and the basic LCA in finding the global solutions of different benchmark problems. The results indicated the better performance of RLCA in terms of the quality of final solutions and the convergence speed. LCA has been applied to different optimization problems.

H. Rezaei · O. Bozorg-Haddad (✉)
Department of Irrigation and Reclamation Engineering, Faculty of Agricultural Engineering
and Technology, College of Agriculture and Natural Resources, University of Tehran, Karaj,
Tehran 31587-77871, Iran
e-mail: OBHaddad@ut.ac.ir

H. Rezaei
e-mail: HosseinRezaie18@ut.ac.ir

X. Chu
Department of Civil and Environmental Engineering, North Dakota State University,
Dept 2470, Fargo, ND 58108-6050, USA
e-mail: Xuefeng.Chu@ndsu.edu

© Springer Nature Singapore Pte Ltd. 2018
O. Bozorg-Haddad (ed.), *Advanced Optimization by Nature-Inspired Algorithms*,
Studies in Computational Intelligence 720, DOI 10.1007/978-981-10-5221-7_3

Lenin et al. (2013) utilized LCA for solving a multi-objective dispatch problem. Abdulhamid and Latiff (2014) used LCA based on job scheduling scheme for optimization of infrastructure as a service cloud. Sajadi et al. (2014) applied LCA for scheduling context. The scheduling considered in a permutation flow-shop system with makespan criterion. Abdulhamid et al. (2015a, b, c) used LCA to minimize makespan time scheduled tasks in IaaS cloud. Abdulhamid et al. (2015a, b, c) proposed a job scheduling algorithm based on the enhanced LCA in order to optimize the infrastructures as a service cloud. The performance of the proposed algorithm was compared with three other algorithms. The results proved that the LCA scheduling algorithm performed better than other algorithms. Xu et al. (2015) presented an improved league championship algorithm with free search (LCAFS), in which the novel match schedule was implemented to improve the capability of competition for teams and by introducing the free search operation, the diversity of league was also improved. The global search performance and convergence speed of LCAFS were compared with those of the basic LCA, PSO, and genetic algorithm (GA) in solving some benchmark functions. The results demonstrated that LCAFS was able to describe complex relationships between key influence factors and performance indexes. Jalili et al. (2016) introduced a new approach for optimizing truss design based on LCA, which considered the concept of tie. The performance of the proposed algorithm was evaluated by using five typical truss design examples with different constraints. The results illustrated the effectiveness and robustness of the proposed algorithm.

3.2 Review of LCA and Its Terminology

The basic idea of LCA was inspired by the concept of league championship in sport competitions. Following terms related to the league, team and its structure are commonly used in LCA.

League: 'league' means a group of sport teams that are organized to compete with each other in a certain type of sport. A league championship can be held in different ways. For example, the number of games that each team should play with other teams may vary. At the end of the league championship, the champion is determined based on the win, loss and tie records during the league's competition with other teams.

Formation: formation of a team refers to the specific structure of the team when playing with other teams, such as the positions of players, and the rule of each player during match. For any sport teams, coaches arrange their teams based on their players' abilities to achieve the best available formation to play with other teams.

Match analysis: match analysis refers to the examination of behavioral events occurring during a match. The main goal of match analysis after determining the performance of a team is to recognize the strengths and weaknesses and to improve. The important part of the match analysis process is to send the feedback of last matches (their own match and the opponent's match) to players. The feedback should be given to the player, pre-match and post-match to build up team for next match. One of these analyses is the strength/weakness/opportunity/threat (SWOT) analysis which links the external (opportunities and threats) and internal (strengths and weaknesses) factors of the team's performance. Identification of SWOTs is necessary because next planning step is based on the results of the SWOT analysis to achieve the main objective. The SWOT analysis evaluates the interrelationships between the internal and external factors of match in four basic categories:

S/T matches illustrate the strengths of a team and the threats of competitors. The team should use its strengths to defuse threats.

S/O matches illustrate the strengths and opportunities of a team. The team should use its strengths to take opportunities.

W/T matches illustrate the weaknesses and threats of a team. The team should try to minimize its weaknesses and defuse threats.

W/O matches show the weaknesses coupled with opportunities of a team. The team should attempt to overcome weaknesses by use of opportunities.

The SWOT analysis provides a structure for conducting gap analysis. A gap refers to the space between the place where we are and the place where we want to be. The identification process of a team's gap contains an in-depth analysis of the factors that express the current condition of the team and subsequently help to make a plan for improvement of the team.

3.3   League Championship Algorithm

LCA is a population-based algorithm that is used for solving global optimization problems with a continuous search space. Like other population-based algorithms, LCA tries to move populations from possible areas to promising areas during searching for the optimum in the whole decision space. Table 3.1 presents a list of the characteristics of the LCA.

Table 3.1 Characteristics of the LCA

General algorithm	League championship algorithm
Decision variable	Player's strength in each team's formation
Solution	Team's formation
Old solution	Old team's formation
New solution	New team's formation
Best solution	The winner of league championship
Fitness function	Team's playing strength
Initial solution	Random formation for each team
Selection	Match analysis process
Process of generating new solution	SWOT matrix

In the optimization process of LCA, a set of L (an even number) solutions are first created randomly to build initial population. Then, they evolve gradually the composition of the population in sequential iterations. In LCA, league refers to population; formation of a team stands for solution; and week refers to iteration. So, team i denotes the ith solution of the population. A fitness value is then calculated for each team based on the team's adaption to the objectives (determined by the concepts of player strength and team's formation). In LCA, new solutions are generated for next week by applying operators to each team based on the results of match analysis which are used by coaches to improve their team's arrangements. An evolutionary algorithm (EA) is a population-based one that uses the Darwin's evolution theory as selection mechanism. Based on a pseudo code of EA and according to the selection process of LCA (greedy selection), in which the current team's formation is replaced by the best team's formation, LCA can be classified as an EA group of the population-based algorithms. LCA terminates after a certain number of seasons (S), each of which is composed of $(L - 1)$ weeks. Note that the number of iterations in LCA is equal to $S(L - 1)$.

LCA models an artificial championship during the optimization process of the algorithm based on some idealized rule that can be expressed as follows:

(1) The team with better playing strength (ability of the team to defeat competitors) has more chances to win the game.
(2) The weaker team can win the game but its chance to win the game is very low (the playing strength does not determine the final outcome of the game exactly).
(3) The sum of the win's probabilities of both teams that participate in a match is equal to one.
(4) The outcome of the game only can be win or loss (tie is not acceptable as the outcome of the game in the basic version of LCA).
(5) When teams i and j compete with each other and eventually team i wins the match, any strength helps team i to win and dual weakness causes team j to lose the match (weakness is a lack of specific strength).
(6) The teams just focus on their forthcoming match without consideration of other future matches and the formation of team arranged only by previous week results.

Figure 3.1 shows the flowchart of LCA, which illustrates the optimization process of the basic LCA. As shown in Fig. 3.1, first of all a representation for individuals must be chosen. Solutions (team's formation) are represented with n decision variables of real numbers. Each element of the solutions depends on one of the team's players and shows the corresponding values of the variables with the aim of optimization. Changes in each value can be the effect of changes in the responsibility of the corresponding player. $f(x_1, x_2, \ldots, x_n)$ denotes an n variable function to be minimized during the optimization running of LCA over a decision space (a subset of R^n). The solution of team i at week t can be represented by $X_i^t = \left(x_{i1}^t, x_{i2}^t, \ldots, x_{in}^t\right)$ and the value of its fitness function (player strength) is $ff(x_i^t)$. $B_i^{t-1} = \left(b_{i1}^{t-1}, b_{i2}^{t-1}, \ldots, b_{in}^{t-1}\right)$ and $ff(B_i^{t-1})$ denote the best formation of team i before

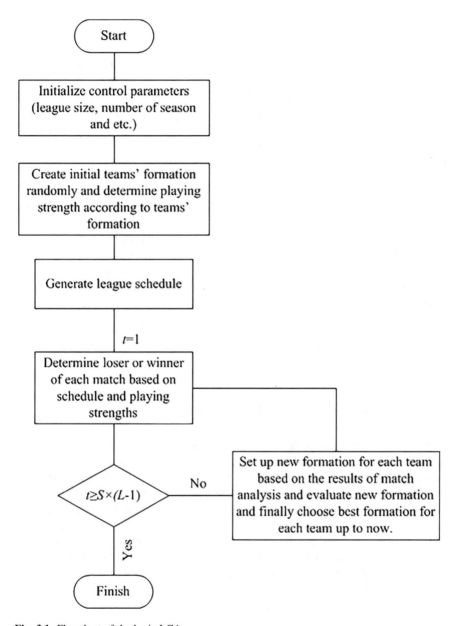

Fig. 3.1 Flowchart of the basic LCA

week t and its fitness function, respectively. The greedy selection in LCA can be made between $ff(x_i^t)$ and $ff(B_i^{t-1})$. The modules of LCA, generation of league schedule, determination of the winner or loser, and setup of new formation are detailed in the following section.

3.4 Generating League Schedule

The common aspect of different sport leagues is the structure, in which teams can compete with each other in a nonrandom schedule, named season. Therefore, the first and the most important step in LCA, is to determine the match schedule in each season. A single-round robin schedule can be applied in LCA to determine the team's schedule, in which each team competes against other teams just once in a certain season. In a championship containing L teams with the single-round robin schedule rule, there are $L(L-1)/2$ matches in a certain season. In each of $(L-1)$ weeks, $(L/2)$ matches will be held in parallel (if L is odd, in each week $(L-1)/2$ matches will be held and one team has to rest).

The procedure of scheduling of the algorithm can be illustrated by a simple example of league championship of 8 teams. The teams have named from a to h. In the first week, the competitors are identified randomly. Figure 3.2a shows the competitors in the first week. For example, team a competes with team d and team b competes with team g. In the second week, in order to identify the pairs of competitors, one of the teams (team a) is fixed in its own place and all other teams turn round clockwise. Figure 3.2b indicates the procedure of identifying the pairs of competitors for week 2. This process continues until the last week (week 7) shown Fig. 3.2c. In LCA, the single-round robin tournament is applied for scheduling L teams in $S(L-1)$ weeks.

3.5 Determining the Winner or Loser

During the league championship, teams compete with each other in every week. The outcome of each match can be loss, tie, or win. The scoring rules for the outcome of the matches can be different for different sports. For instance, in soccer the winner gets three, and the loser gets a zero score. By the end of the match, both teams get one if the outcome is tie. According to the idealized rule 1, the chance of a stronger team to win the match is higher than its competitor, but occasionally a weaker team may win the match. Therefore, the outcome of the match is associated with different reasons. The most important one is the playing strengths of the teams. So we can consider a linear relationship between the playing strengths and the outcome of the match (idealized rule 2).

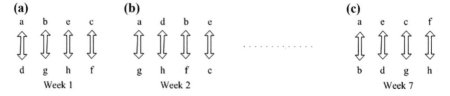

Fig. 3.2 A simple example of league championship scheduling

In LCA, a stochastic criterion of playing strength, which depends on the degree of fit of a team, is utilized to determine the winner or loser of the match. Note that in the basic version of LCA the outcome of the matches can only be win or loss (no tie). The teams' degree of fit refers to the proportion of playing strength, which is calculated based on the distance between the playing strength and the ideal reference point (the lower bound of the optimization problem).

Assuming that team i competes with team j at week t, the chance of each team to defeat another team can be expressed as (Kashan 2009) follows:

$$\frac{f(x_i^t) - \hat{f}}{f(x_j^t) - \hat{f}} = \frac{p_j^t}{p_i^t} \tag{3.1}$$

where x_i^t and x_j^t = formation of teams i and j at week t; $f(x_i^t)$ and $f(x_j^t)$ = playing strength of teams i and j at week t; $\hat{f}$ = ideal reference point; p_i^t = chance of team i to defeat team j at week t; and p_j^t = chance of team j to defeat team i at week t.

Because the chances of both teams to win the match are evaluated based on the specific point, the ratio of distance is identified as the team's winning portion. According to idealized rule 3, the relationship of the chances of teams i and j at week t can be expressed as follows:

$$p_i^t + p_j^t = 1 \tag{3.2}$$

Based on Eqs. (3.1) and (3.2), the chance of team i to defeat team j at week t is given by (Kashan 2014):

$$p_i^t = \frac{f\left(x_j^t\right) - \hat{f}}{f(x_i^t) + f\left(x_j^{t)}\right) - 2\hat{f}} \tag{3.3}$$

In LCA, in order to specify the winner of the match, a random number between [0,1] is generated randomly. If the generated number is equal to or less than p_i^t, team i defeats team j at week t. Otherwise, team j defeats team i at week t.

3.6 Setting Up a New Team Formation

Before applying any strategy to team i, in order to change the formation of team i at next week, coaches should identify strengths and weaknesses of the team and players (individuals). Based on these strengths and weaknesses, coaches determine the formation of the team in next week to enhance the performance of the team. An artificial match analysis can be performed to specify the opportunities and threats. Strengths and weaknesses are internal factors while opportunities and threats are external factors. In LCA, the internal factors are evaluated based on the team's

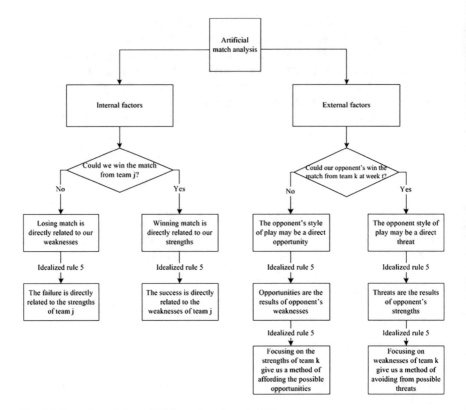

Fig. 3.3 Procedure of the artificial match analysis in LCA

performance at the last week (week t), while evaluating the external factors is based on the opponent's performance at week t. The artificial match analysis helps prepare team i for next week (week $t + 1$). In the modeling process, if team i wins (loses) the match at week t, it is assumed that the success (failure) is directly related to the strengths (weaknesses) of team i or weaknesses (strengths) of its opponent team j (idealized rule 5). The procedure of modeling and evaluating the artificial match analysis for team i at week t is displayed in Fig. 3.3. The left side in Fig. 3.3 shows the evaluation of hypothetical internal factors and the right side shows the way of evaluating the external factors.

According to the results of the artificial match analysis applied to team i in order to determine its performance, the coach should take some possible actions to improve the team's performance. The possible actions (SWOT analysis) are shown in Table 3.2. The SWOT analysis is adjusted based on idealized rule 6. Table 3.2 shows different strategies (S/T, S/O, W/T, and W/O) that can be adopted for team i in different situations. For instances, if team i has won the last match and team l has lost its match at the last week, it is reasonable for team i to focus on strengths which give it more chance to win team l at next match. Therefore, adopting the S/O strategy for team i is efficient. Table 3.2 also displays,

Table 3.2 Hypothetical SWOT analysis derived from the artificial match analysis

	Adopt S/T strategy	Adopt S/O strategy	Adopt W/T strategy	Adopt W/O strategy
	Team i has won	Team i has won	Team i has lost	Team i has lost
	Team l has won	Team l has lost	Team l has won	Team l has lost
	Focusing on	Focusing on	Focusing on	focusing on
S	Own strengths or weaknesses of team j	Own strengths or weaknesses of team j	–	–
W	–	–	Own weaknesses or strengths of team j	Own weaknesses or strengths of team j
O	–	Weaknesses of team l or strengths of team k	–	Weaknesses of team l or strengths of team k
T	Strengths of team l or weaknesses of team k	–	Strengths of team l or weaknesses of team k	–

in a metaphorical way, the SWOT analysis matrix which is used for planning in the future matches.

The aforementioned analysis must be performed by all participants at week t to plan for next match and choose a suitable formation for upcoming match. After adopting a suitable strategy for team i based on the SWOT matrix, all teams should fill their gaps. For instance, assume that in a soccer match team i has lost the match at week t to team j and the results of the match analysis process have specified that the type of defensive state (man-to-man defensive state) is the reason of loss Therefore, a gap exists between the current sensitive defensive state and the state which ensures a man-to-man pressure defense at week $t + 1$.

According to the league schedule, team l is the competitor of team i ($i = 1, 2, ..., L$) at week $t + 1$; team j is the competitor of team i at week t; and team k is the competitor of team l at week t. As aforementioned, X_i^t, X_j^t and X_k^t, respectively, denote the formations of teams i, j, and k at week t. $(X_k^t - X_i^t)$ defines the gap between playing styles of teams i and k, which highlights the strengths of team k. This case applies when team i wants to play with team l at week $t + 1$ and team k wins team l at week t by X_k^t's formation. Therefore, if team i uses the playing style of team k at week t (X_k^t) to compete with team l at week $t + 1$, it is highly possible for team i to win team l at week $t + 1$. Similarly, $(X_i^t - X_k^t)$ is used if we want to 'focus on the weaknesses of team k'. In this case, team i should not use the playing style of team k at week t against team l. $(X_i^t - X_j^t)$ and $(X_j^t - X_i^t)$ also can be defined. Due to the principle that each team should play with the best formation that is selected from playing experience up to now and by considering the results of the artificial match analysis in last week, new formation of team i at week $t + 1$ $\left[(X_i^{t+1} = (x_{i1}^{t+1}, x_{i2}^{t+1}, ..., x_{in}^{t+1}))\right]$ can be set up by one of the following equations:

If teams i and l have won the match at week t, the new formation of team i will be generated based on the S/T strategy:

$$(\text{S/T strategy}) : x_{im}^{t+1} = b_{im}^t + y_{im}^t \left(\omega_1 r_{1im} (x_{im}^t - x_{km}^t) + \omega_1 r_{2im} (x_{im}^t - x_{jm}^t) \right) \qquad (3.4)$$
$$\forall m = 1, 2, \ldots, n$$

If team i has won and team l has lost at week t, the new formation of team i will be generated based on the S/O strategy:

$$(\text{S/O strategy}) : x_{im}^{t+1} = b_{im}^t + y_{im}^t \left(\omega_2 r_{1im} (x_{Km}^t - x_{im}^t) + \omega_1 r_{2im} (x_{im}^t - x_{jm}^t) \right) \qquad (3.5)$$
$$\forall m = 1, 2, \ldots, n$$

If team i has lost and team l has won at week t, the new formation of team i will be generated based on the W/T strategy:

$$(\text{W/T strategy}) : x_{im}^{t+1} = b_{im}^t + y_{im}^t \left(\omega_1 r_{1im} (x_{im}^t - x_{km}^t) + \omega_2 r_{2im} (x_{jm}^t - x_{im}^t) \right) \qquad (3.6)$$
$$\forall m = 1, 2, \ldots, n$$

If teams i and l have lost the match at week t, the new formation of team i will be generated based on the W/O strategy:

$$(\text{W/O strategy}) : x_{im}^{t+1} = b_{im}^t + y_{im}^t \left(\omega_2 r_{1im} (x_{km}^t - x_{im}^t) + \omega_2 r_{2im} (x_{jm}^t - x_{im}^t) \right) \qquad (3.7)$$
$$\forall m = 1, 2, \ldots, n$$

where m = number of team members; r_{1im} and r_{2im} = random numbers between [0,1]; ω_1 and ω_2 = coefficients used to scale the contribution of approach or retreat components; and y_{im}^t = binary variable that specifies whether or not the mth player must change in the new formation (only $y_{im}^t = 1$ allows to change). Note that different signs in the parentheses are the consequence of acceleration towards the winner or recess from the loser. $Y_i^t = (y_{i1}^t, y_{i2}^t, \ldots, y_{in}^t)$ denotes a binary change variable. The summation of the changes needed for next match ($y_i^t = 1$) is equal to q_i^t. Changes in all aspects of the team (players and styles) by coaches are not common (just a few changes in the team can be required). In order to calculate the number of changes in the team's formation for next match, a truncated geometric distribution is applied in LCA. The truncated geometric distribution lets LCA to control the number of changes with emphasis on the smaller rates of changes in B_i^t. The truncated geometric distribution can be expressed as follows:

$$q_i^t = \left[\frac{\ln(1 - (1 - p_c)^{n - q_0 + 1}) r}{\ln(1 - p_c)} \right] + q_0 - 1 : q_i^t \in \{q_0, q_0 + 1, \ldots, q_0 + n\} \qquad (3.8)$$

where r = random number between [0,1]; p_C = control parameter $[p_c < 1, \ p_c \neq 0]$; and q_0 = the least number of changes. If $p_c < 0$, the situation is reversed so that by a more negative value of p_c, The emphasis in LCA is placed on a greater rate of change in the team's formation. q_0 is determined during the match analysis (note that the minimum value of q_0 is equal to zero). p_c in the truncated geometric distribution is the probability of success. In LCA, after calculating the value of q_i^t by using Eq. (3.8), the players of B_i^t are randomly selected and changed based on Eqs. (3.4)–(3.7).

3.7 Pseudo Code of LCA

Begin

Generate initial teams' formation randomly [$(X_i^t = (x_{i1}^t, x_{i2}^t,...,x_{in}^t)$], $\forall i = 1, 2, ...,L]$

Generate league schedule for L teams

For m=1: $L(S-1)$

Evaluate the strengths for all teams

Calculate the chance of each team to defeat its competitors in next match (p_i^t)

Generate a random number between [0,1] (R_n)

If $R_n \le p_i^t$

Team i wins the match

Else

Team j wins the match

End if

Generate a random number between [0,1] (r)

Calculate the number of changes in teams' best formation (B_i^t) for next match

based on the truncated geometric distribution (q_i^t)

q_i^t players are selected randomly from B_i^t and changed by the SWOT matrix

If team i and team l have won

Select the S/T strategy

Else if team i has won and team l has lost

Select the S/O strategy

Else if team i has lost and team l has won

Select the W/T strategy

Else if team i has lost and team l has lost

Select the W/O strategy

End if

End for m

End

3.8   Conclusions

This chapter described the league championship algorithm (LCA), which stemmed from the concept of league championship in sport. This chapter also presented a literature review of LCA, and its algorithmic fundamental and pseudo code.

References

Abdulhamid, S. M., & Latiff, S. A. (2014). *League championship algorithm (LCA) based job scheduling scheme for infrastructure as a service cloud.* 5th International Graduate Conference on Engineering, Science and Humanities, UTM Postgraduate Student Societies, Johor, Malaysia, 19–21 August.

Abdulhamid, S. M., Latiff, M. S., & Abdullahi, M. (2015a). *Job scheduling technique for infrastructure as a service cloud using an enhanced championship algorithm.* 2nd International Conference on Advanced Data and Information Engineering, Lecture Notes in Electrical Engineering, Bali, Indonesia, 25–26 April.

Abdulhamid, S. M., Latiff, M. S., & Idris, I. (2015b). Tasks scheduling technique using league championship algorithm for makespan minimization in IaaS cloud. *ARPN Journal of Engineering and Applied Sciences., 9*(12), 2528–2533.

Abdulhamid, S. M., Lattif, M. S. A., Madni, S. H. H., & Oluwafemi, O. (2015c). A survey of league championship algorithm: prospects and challenges. *Indian Journal of Science and Technology.*

Jalili, S., Kashan, A. H., & Husseinzadeh, Y. (2016). League championship algorithms for optimization design of pin-jointed structures. *Journal of Computing in Civil Engineering.* doi:10.1061/(ASCE)CP.1943-5487.0000617

Kashan, A. H. (2009). *League championship algorithm: A new algorithm for numerical function optimization.* International Conference on Soft Computing and Pattern Recognition, IEEE Computer Society, Malacca, Malaysia, 4–7 December.

Kashan, A. H. (2011). An efficient algorithm for constrained global optimization and application to mechanical engineering design: League championship algorithm (LCA). *Computer-Aided Design, 43*(2011), 1769–1792.

Kashan, A. H. (2014). League championship algorithm (LCA): an algorithm for global optimization inspired by sport championships. *Applied Soft Computing, 16*(2014), 171–200.

Kashn, A. H., Karimiyan, S., Karimiyan, M., & Kashan, M. H. (2012). *A modified League Championship Algorithm for numerical function optimization via artificial modeling of the "between two halves analysis".* The 6th International Conference on Soft Computing and Intelligent Systems, and the 13th International Symposium on Advanced Intelligence Systems, University of Aizu, Kobe, Japan, 20–24 November.

Lenin, K., Reddy, B. R., & Kalavati, M. S. (2013). League championship algorithm (LCA) for solving optimal reactive power dispatch problem. *International Journal of Computer and Information Technologies., 1*(3), 254–272.

Sajadi, S. M., Kashan, A. H., & Khaledan, S. (2014). A new approach for permutation flow-shop scheduling problem using league championship algorithm. In *Proceedings of CIE44 and IMSS'14, 2014.*

Xu, W., Wang, R., & Yang, J. (2015). An improved league championship algorithm with free search and its application on production scheduling. *Journal of Intelligent Manufacturing.* doi:10.1007/s10845-015-1099-4

Chapter 4
Anarchic Society Optimization (ASO) Algorithm

Atiyeh Bozorgi, Omid Bozorg-Haddad and Xuefeng Chu

Abstract Due to limited resources and equipment in most engineering projects, it is necessary to use optimization techniques. Older optimization techniques, including derivative and other mathematical methods may not be practical to new complex problems. Therefore new optimization algorithms are needed. In the past decades many algorithms were developed and used for different optimization problems, which can be divided into three categories including classic, evolutionary and heuristic algorithms. The evolutionary and heuristic algorithms which are used widely in recent years are based on animals' life. In this chapter, one of the heuristic algorithms named Anarchic Society Optimization (ASO) algorithm based on human societies, is introduced. After a brief literature review of the ASO algorithm, more technical details on this method and its performance are described.

4.1 Introduction

Anarchic is derived from the Greek word *anarkos* meaning "no boss" and Anarchia means "lack of government". The term Anarchism refers to a political opinion and movement believing that any political power and authority are obscene and unnecessary and that any government should be overthrown and replaced with free associations and volunteer groups. Because the Anarchism believes that the government causes a nation's social miseries. Overall, Anarchists are opposed to any

A. Bozorgi · O. Bozorg-Haddad (✉)
Department of Irrigation and Reclamation Engineering, Faculty of Agricultural Engineering and Technology, College of Agriculture and Natural Resources, University of Tehran, 31587-77871 Karaj, Tehran, Iran
e-mail: OBHaddad@ut.ac.ir

A. Bozorgi
e-mail: Bozorgi.Atiyeh@ut.ac.ir

X. Chu
Department of Civil and Environmental Engineering, North Dakota State University, Dept 2470, Fargo, ND 58108-6050, USA
e-mail: Xuefeng.Chu@ndsu.edu

© Springer Nature Singapore Pte Ltd. 2018   31
O. Bozorg-Haddad (ed.), *Advanced Optimization by Nature-Inspired Algorithms*,
Studies in Computational Intelligence 720, DOI 10.1007/978-981-10-5221-7_4

government authority and consider the democracy as the Tyranny of the majority. They emphasize individual freedom. This emphasis results in opposition to any external authority, especially government, which is construed as a barrier for free individual growth and excellence.

The Anarchist thought is based on a variety of principles including individualism, humanism, libertarians, lawlessness, anarchic, and absolute freedom. According to these principles, Anarchism opposes any religious or non-religious social institutions and considers the human as an absolute free creature. At the heart of Anarchism, there is a reckless utopia orientation, believing in natural wellness, or at least mankind potential wellness (Nettlau 2000).

Anarchism contains a variety of branches and anarchism theorists follow one of them. According to the view of the Communist Anarchism, human is inherently social, and a society and its individuals benefit each other. Human and society conformity is possible when negating the powerful social institutions especially government. The Syndicate-oriented Anarchism looks for the salvation in economic strife not in the political strife of proletariat. Followers of this faction organize labor unions and syndicates to quarrel with the power structure. According to their point of view, the current government will be eventually annihilated as a result of a revolution and the new economic order will be formed based on syndicates. Nowadays, such thoughts have become a mass movement in some South American and European countries. The followers of Individualist Anarchism believe that the human has the right to do whatever he/she will and whatever deprive him/her from freedom must be destroyed (Woodcook 2004).

In general, there are three major insufficiencies that Anarchism suffers from. First, Anarchism has the unrealistic goal that is related to collapse of government and all other forms of political authority, while economic and social development has been always accompanied with government roles. Second, the Anarchism objects to powerful institutions like parties that can play an effective and efficient role in development of a society. Third, Anarchism lacks a series of distinct and coherent political beliefs, which causes many disputes.

During the recent centuries, many countries have undergone certain types of anarchy, including France (The revolution period), Jamaica (1720), Russia (during Civil Wars), Spain (1936), Albania (1997), and Somali (1991–2006). According to the view of Anarchists, a society can be managed without the need of the central government and only based on individuals or volunteer groups. In this case, individuals or groups will be able to determine the right direction without being ordered by a ruling power and only based on their or others' previous experiences. Although this viewpoint has not been prosperous in stable management of a society so far, it can be used as a basis for developing optimization methods in engineering sciences.

In this method, each individual selects his/her next position according to the personal experiences, group or syndicate experiences, and historical experiences. Finally, after a number of moves, at least one of the group members would reach a near-optimal answer. Employing this algorithm causes the total decision space to be fully searched and prevents being stuck at local optima.

The ASO algorithm was first introduced by Ahmadi-Javid (2011). He compared the answers obtained from the ASO algorithm with those from the Genetic Algorithm (GA) and the particle swarm optimization (PSO) algorithm and proved that the ASO algorithm resulted in better answers. The author also claimed that the anarchy community algorithm converged to the global optimum with the probability of 100%. Shayeghi and Dadashpour (2012) compared ASO with PSO, vector evaluated particle swarm optimization (VEPSO) and craziness based particle swarm optimization (CRPSO) algorithm for voltage oscillation damping problem and the best results were obtained from the ASO algorithm. Ahmadi-Javid and Hooshangi-Tabrizi (2012) expanded the ASO algorithm to two objective functions and the comparison between ASO and PSO algorithms showed that ASO provided much better results. Ahmadi Javid and Hooshangi-Tabrizi (2015) expanded and applied the ASO algorithm for permutation flow-shop scheduling problem with integer and linear objective functions namely ASO(I) and ASO(II). The numerical results obtained showed that the ASO algorithm had higher efficiency for that problem.

4.2 Formulation

The detailed procedures of the ASO algorithm are shown in Fig. 4.1.

For a solution space S, $f : S \rightarrow R$ is a function that should be minimized in S. Consider that a community, consisting of N member(s), is being searched in an

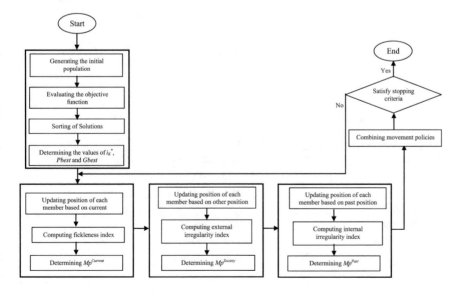

Fig. 4.1 Flowchart of the ASO algorithm

unknown territory (the solution space) for discovering the best place to live (i.e., the overall minimum of f on S). $X_i(k)$ presents the position of the ith member in the kth iteration; $X^*(k)$ denotes the best position experienced by all members in the kth iteration; and G^{best} is the best position experienced by the ith member during the first k iterations.

4.3   Algorithm Procedure

As shown in Fig. 4.1, first, a number of community members are selected randomly within the solution space. Then, the fitness of every member is determined. According to the calculated fitness value and comparison with $X^*(k)$, P_i^{best} and G^{best}, the movement policy and a new position of the member will be determined. After an adequate number of iterations, at least one of the members will reach the optimal position. Table 4.1 lists the characteristics of the ASO.

4.4   Movement Policy Based on Current Positions

The first movement policy for the ith member in the kth iteration $[MP_i^{current}(k)]$ is adopted based on the current position. The Fickleness Index $FI_i(k)$ for member i in iteration k is used to select this movement policy (Ahmadi-Javid 2011). This index measures the satisfaction of the current position of the ith member compared with other members' positions. If the objective function is positive in S, the Fickleness Index can be expressed as one of the following forms (Ahmadi-Javid 2011):

$$FI_i(k) = 1 - \alpha_i \frac{f(X^*(k))}{f(X_i(k))} - (1 - \alpha_i)\frac{f(P_i(k))}{f(X_i(k))} \tag{4.1}$$

Table 4.1 Characteristics of the ASO algorithm

General algorithm	Anarchic society optimization
Decision variable	Society members' positions in each dimension
Solution	Society members' positions
Old solution	Old positions of society members
New solution	New positions of society members
Best solution	Best position
Fitness function	Desirability of members' positions
Initial solution	Random positions
Selection	–
Process of generating new solution	Combination of movement policy

$$FI_i(k) = 1 - \alpha_i \frac{f(G(k))}{f(X_i(k))} - (1 - \alpha_i) \frac{f(P_i(k))}{f(X_i(k))} \tag{4.2}$$

where α_i is a non-negative number in $[0,1]$. Thus, the Fickleness Index is a number in the range of $[0,1]$. According to the values of Fickleness Index, the ith member would select his/her next position. If $FI_i(k)$ is smallest, the ith member has the best position among all members. So it is better to select the movement policy based on $X^*(k)$. Otherwise, the ith member has a totally unpredictable movement. Therefore the movement policy for the ith member according to the value of $FI_i(k)$ can be expressed as:

$$MP_i^{\text{current}}(k) = \begin{cases} \text{moving towards } X^*(k) & 0 \leq FI_i(k) \leq \alpha_i \\ \text{moving towards a random } X_i(k) & \alpha_i \leq FI_i(k) \leq 1 \end{cases} \tag{4.3}$$

4.5   Movement Policy Based on Positions of Other Members

The second movement policy for the ith member in the kth iteration $[MP_i^{\text{society}}(k)]$ is adopted based on the positions of other members. Although each member should move in the direction of G^{best} logically, the movement of the member is not predictable due to the anarchist nature of the member and may move toward another community member. Therefore, the external irregularity index $EI_i(k)$ for the ith member in the kth iteration can be calculated by (Ahmadi-Javid 2011):

$$EI_i(k) = 1 - e^{-\theta_i \, [f(X_i(k)) - f(G(k))]} \tag{4.4}$$

$$EI_i(k) = 1 - e^{-\delta_i \, D(k)} \tag{4.5}$$

in which θ_i and δ_i are positive numbers and $D(k)$ is an appropriate dispersion measure like coefficient of variation $CV(k)$. Equation (4.4) defines the distance of community member i from G^{best}. If the community member is close to G^{best}, it will have a more logic behavior. Otherwise, it shows an anarchic behavior based on anarchy. Equation (4.5) defines a diversity index in community which has a direct relationship with the diversity of the community members. In the case that this index is selected, the community members are supposed to behave more logically and they are less diversified. Thus, with consideration of a threshold for $EI_i(k)$, it is possible to define the movement policy based on the positions of other members as follows:

$$MP_i^{\text{society}}(k) = \begin{cases} \text{moving towards } G^{\text{best}} & 0 \leq EI_i(k) \leq \text{threshold} \\ \text{moving towards a random } X_i(k) & \text{thereshold} \leq EI_i(k) \leq 1 \end{cases} \quad (4.6)$$

The closer the threshold is to zero, the more illogical the member movements would be. As the threshold converges to one, the members would behave more logically.

4.6 Movement Policy Based on Previous Positions

The third movement policy for the ith member in the kth iteration [$MP_i^{\text{past}}(k)$] is adopted based on the previous positions of the individual member. In order to select this movement policy, the position of the ith member in the kth iteration is compared to P_i^{best}. If the position of the member is close to P_i^{best}, the member behaves more logically. Otherwise, the member shows illogical behavior. To determine the movement policy based on previous positions, the internal irregularity index $II_i(k)$ for the ith member in the kth iteration is defined as follows:

$$II_i(k) = 1 - e^{-\beta_i [f(X_i(k)) - f(P_i(k))]} \quad (4.7)$$

where β_i is a positive number. Like the previous policy, with selection of a threshold for $II_i(k)$, the movement policy can be defined based on previous positions as follows:

$$MP_i^{\text{past}}(k) = \begin{cases} \text{moving towards } P_i^{\text{best}} & 0 \leq II_i(k) \leq \text{threshold} \\ \text{moving towards a random } X_i(k) & \text{thereshold} \leq II_i(k) \leq 1 \end{cases} \quad (4.8)$$

The closer the threshold is to zero, the more illogical the member movements would be. As the threshold converges to one, the members would behave more logically.

4.7 Combination of Movement Policies

In order to select the final movement policy, the three policies discussed above are combined with each other. After the movement policies are calculated, each member should combine these policies with a method and move toward a new position. One of the simplest methods is to select the policy with the best answer. The next alternative is to combine the movement policies with each other sequentially which is referred to as the sequential combination rule. The crossover method can either be used for continuous problems coded as chromosomes, or used in a sequential way to combine the movement policies. Ahmadi Javid (2011) demonstrated that the ASO algorithm is a more general state of the PSO algorithm.

4.8   Pseudo Code of the ASO

Begin

      Input the parameters of the algorithm and initial data

      Generate M initial solutions and evaluate their fitness values

      While (termination criteria are not satisfied)

      Determining the values of $X^*(k)$, P_i^{best} and G^{best}

          For i=1:M

          Computing $FI_i(k)$

               If $FI_i(k)$ is less than threshold $X_i(k)$, then $X_i(k)$ moves towards $X^*(k)$

               Else $X_i(k)$ moves towards a random member

               End if

          End for

          For i=1:M

          Computing $EI_i(k)$

               If $EI_i(k)$ is less than threshold $X_i(k)$, then $X_i(k)$ moves towards G^{best}

               Else $X_i(k)$ moves towards a random member

               End if

          End for

          For i=1:M

          Computing $II_i(k)$

               If $II_i(k)$ is less than threshold $X_i(k)$, then $X_i(k)$ moves towards P_i^{best}

               Else $X_i(k)$ moves towards a random member

               End if

          End for

          For i=1:M

          Updating the position by combining the movement policies

          Calculating the fitness values for the members of society

          End for

      End while

      Report the best solution

End

4.9 Conclusion

The evolutionary and heuristic algorithms are widely used for solving optimization problems in the engineering and science fields. In the last few decades, various algorithms have been introduced mostly based on insects, animal lives, and biological concepts. In this chapter, the Anarchic Society Optimization algorithm, first introduced by Ahmadi-Javid (2011), was described.

This algorithm has been investigated for solving electrical and industrial engineering problems (e.g., power) and optimizing water networks and reservoir operation. The ASO algorithm has some advantages such as its relatively simple structure and its potential to achieve better performance. In definition of the ASO algorithm three indices are used. It seems that changing the use of these indices to reach a new position for each member or even defining a new index can lead to superior convergence.

The ASO algorithm was adopted from the life of human communities. Since human societies are more complicated than animal or insect groupings, it is expected that ASO as the first algorithm based on human societies is a turning point of the performance and capabilities of optimization algorithms.

References

Ahmadi-Javid, A. (2011). Anarchic Society Optimization: A human-inspired method. In *IEEE Congress on Evolutionary Computation (CEC)* (pp. 2586–2592), New Orleans, LA.

Ahmadi-Javid, A., & Hooshangi-Tabrizi, P. (2012). *An Anarchic Society Optimization Algorithm for a flow-shop scheduling problem with multiple transporters between successive machines.* International Conference on Industrial Engineering and Operations Management (ICIEOM), Istanbul, Turkey, 3–6 July.

Ahmadi-Javid, A., & Hooshangi-Tabrizi, P. (2015). A mathematical formulation and anarchic society optimisation algorithms for integrated scheduling of processing and transportation operations in a flow-shop environment. *International Journal of Production Research, 53*(19), 5988–6006.

Nettlau, M. (2000). *A short history of anarchism* (1st ed.). Freedom Press. ISBN-13: 978-0900384899.

Shayeghi, H., & Dadashpour, J. (2012). Anarchic society optimization based PID control of an Automatic Voltage Regulator (AVR) system. *Electrical and Electronic Engineering, 2*(4), 199–207.

Woodcook, G. (2004). *Anarchism*. Toronto: Higher Education Division, University of Toronto Press. ISBN 13: 978-1551116297.

Chapter 5
Cuckoo Optimization Algorithm (COA)

Saba Jafari, Omid Bozorg-Haddad and Xuefeng Chu

Abstract The cuckoo optimization algorithm (COA) is used for continuous non-linear optimization. COA is inspired by the life style of a family of birds called cuckoo. These birds' life style, egg laying features, and breeding are the basis of the development of this optimization algorithm. Like other evolutionary approaches, COA is started by an initial population. There are two types of the population of cuckoos in different societies: mature cuckoos and eggs. The basis of the algorithm is made by the attempt to survive. While competing for being survived, some of them are demised. The survived cuckoos immigrate to better areas and start reproducing and laying eggs. Finally, the survived cuckoos are converged in a way that there is a cuckoo society with the same profit rate.

5.1 Introduction

Rajabioun (2011) proposed a new evolutionary algorithm called cuckoo optimization algorithm (COA), which was inspired by the life of cuckoos. Kahramanli (2012) developed a modified cuckoo optimization algorithm (MCOA) and used it to solve two constrained continuous engineering optimization problems. The results showed that MCOA was a powerful optimization method that yielded better solutions to engineering problems. Rabiee and Sajedi (2013) used COA for solving

S. Jafari · O. Bozorg-Haddad (✉)
Department of Irrigation and Reclamation Engineering, Faculty of Agricultural Engineering and Technology, College of Agriculture and Natural Resources, University of Tehran, Karaj, Tehran 31587-77871, Iran
e-mail: OBHaddad@ut.ac.ir

S. Jafari
e-mail: Saba.Jafari@ut.ac.ir

X. Chu
Department of Civil and Environmental Engineering, North Dakota State University, Dept 2470, Fargo, ND 58108-6050, USA
e-mail: Xuefeng.Chu@ndsu.edu

© Springer Nature Singapore Pte Ltd. 2018
O. Bozorg-Haddad (ed.), *Advanced Optimization by Nature-Inspired Algorithms*,
Studies in Computational Intelligence 720, DOI 10.1007/978-981-10-5221-7_5

job scheduling in a grid computational design. The disadvantages of the evolutionary techniques such as genetic algorithm (GA), simulated annealing (SA), particle swarm optimization (PSO), and ant colony optimization (ACO) to solve the problems in the grid schedule, are early convergence and trapping in local optima in large-scale problems. Their results showed that the proposed method was more efficient and had better performance than GA and PSO. Balochian and Ebrahimi (2013) optimized parameters for Sugeno-type fuzzy logic controllers (S-FLCs) that were applied for liquid level control. A programmable logic controller (PLC) was used with fuzzy controller. The results indicated that the optimized FLC by COA had better performance than the one with manual setting of the system parameters for different datasets. Khajeh and Jahanbin (2014) developed a solid phase extraction method using a new sorbent (zinc oxide nanoparticles-chitosan) for pre-concentration and determination of uranium from water samples. A coupled cuckoo optimization algorithm–artificial neural network (COA–ANN) model was developed for simulation and optimization. The optimum limit of detection and the enrichment factor of uranium were 0.5 μg L^{-1} and 125, respectively. Mellal and Williams (2015a) studied the multipass turning process, one of the widely used machining methods in manufacturing industry. They considered minimization of the unit production cost as a key objective of the operation. In their work, cutting parameters were optimized by using COA. The results showed that COA was better than a wide range of other optimization algorithms. Singh and Rattan (2014) employed the COA for the optimization of linear and non-uniform circular antenna arrays. COA was used to ascertain a set of parameters of antenna elements that provided a required radiation pattern. The effectiveness of COA for design of antenna arrays was confirmed by their numerical results. The results showed the superior performance of COA compared to other methods for designing linear and circular antenna arrays. Shadkam and Bijari (2014) evaluated the performance of COA with the Rastrigin function, a continuous non-linear function that was used for evaluating optimization algorithms. The aforementioned function was solved with artificial bee colony (ABC) and the firefly algorithm (FA). Comparison of the results showed that COA had better performance than other algorithms. Shokri-Ghaleh and Alfi (2014) designed an optimal controller for synchronization of bilateral teleoperation systems with the aim of reducing factors such as time delay in communication channels and modeling uncertainties. A novel meta-heuristic algorithm named COA was used. Comparative simulations were performed to determine the feasibility of the proposed control method. The results showed that CO yielder better solutions to the problem than other algorithms including biogeography-based optimization, imperialist competition, and artificial bee colony. Khajeh and Golzary (2014) developed the zinc oxide nanoparticles-chitosan based on extraction of the solid phase for separation and pre-concentration of a trace amount of methyl orange from water samples. The COA-ANN model was used for optimization and simulation. The optimum limit of detections factor of methyl orange was 0.7 μg L^{-1}. Mellal and Williams (2015b) studied a complex engineering optimization problem called CHPED (combined heat and power economic dispatch). The aim was to minimize the system production costs by taking different constraints into consideration. In their study, COA was employed by a penalty function (PFCOA) to solve the CHPED problem. Two case studies of CHPED were presented and the results were compared with those obtained by several other optimization methods, which proved the superior performance of PFCOA. Moezi et al. (2015) determined the location and depth of a crack by measuring its natural frequency changes and used the MCOA numerical method for open edge-crack detection in an Euler–Bernoulli cantilever beam. The crack was modeled by a torsional spring and the coefficient was calculated by using the crack dimensions. The objective function was the weighted squared difference between the calculated and measured natural frequencies. The results of the numerical simulations and experimental tests showed the high accuracy in finding the location and depth of the crack. Mellal and Williams (2016) considered parameter optimization of advanced machining processes to produce complex profiles and high quality products. COA and the Hoopoe heuristic algorithm were used for optimization of the parameters for two conventional machining processes (drilling process and grinding process) and four advanced machining processes (abrasive jet machining, abrasive water jet machining, ultrasonic machining, and water jet machining). Finally, the results were compared with those from other optimization algorithms.

5.2 Cuckoo Life Style

All the 9000 existing birds in the world have the same reproduction way: egg laying. None of the birds give birth. They lay eggs and raise the baby birds outside their bodies. The larger the eggs are, the less the probability is for a female bird to have more than one egg in her body simultaneously, because on one hand, bigger eggs make flying tough and require more energy to fly. On the other hand, eggs are a rich source of protein for the predators, so it is necessary for birds to find a secure place for egg laying and hatching their eggs. Finding a secure place for egg laying, hatching, and raising the birds until being independent of their parents is of vital importance, which is intellectually solved by birds. They use an artistry and a complicated engineering to do this. The variety of nest-making and the architecture of the nests are absolutely stunning. Most birds make their nests segregated and hidden to prevent being detected by the predators. Some of them hide their nests so skillfully that human beings are not able to recognize and see them.

There are some birds that detached themselves from the challenge of nest-making and use a cunning way to raise their families. These birds are called "Brood Parasites" that never build a nest. They lay their eggs in other species' nests and wait for them to take care of their young. Cuckoo is the most famous "Brood Parasite" that is an expert in deceiving cruelly. The strategy of cuckoos includes their speed, being stealthy and surprising. A mother cuckoo destroys the host's eggs and lay her own eggs among others in the nest and flies away from the location fast and lays caring on the host bird. This process is hardly more than 10 s. Cuckoos make other nests parasitized by their eggs and mimic the color and the patterns of existing eggs carefully so that new eggs in the nest look like the previous eggs. Each female cuckoo is specialized on specific species of birds. This is one of the main secrets of nature about how female cuckoos imitate a special kind of other birds' eggs accurately. Some of the birds recognize cuckoos' eggs and sometimes they even throw the eggs out of the nest. Some of them completely leave the nest and build a new one. In fact, cuckoos continuously improve their mimicry from the eggs in the target nests and host birds learn new ways to recognize the strange eggs as well. This struggle for survival among different birds and cuckoos is a constant and continuous process.

A suitable habitat for cuckoos should provide food sources (specially insects) and locations for laying eggs, so the main necessity of brood parasites will be the habitats of the host species. Cuckoos are found in a variety of places. Most of the species are found in forests, especially evergreen rain forests. Some of the cuckoo species select a wider range of places to live, which can even include dry areas and deserts. Immigrant species select vast environments to make maximum misuse of the host birds. Most of the cuckoo species are non-immigrant but there are several ones that have seasonal immigration as well. Some of the species have partial immigrations in their habitat range. Some species (e.g., channel-billed cuckoos) have diurnal immigration, while others (e.g., yellow-billed cuckoo) have nocturnal immigration. For those cuckoos that live in mountainous areas, availability of the foods necessitates to immigrate to tropical areas. Long-tailed Koel cuckoos which live and lay eggs in New Zealand, immigrate to Micrones, Melanesia, and Polynesia in winters. Yellow-billed species and black-billed species that breed in North America, pass the Caribbean Sea in a non-stop 4000-km flight. Other long-distance immigrations include lesser cuckoos that fly over Indian Ocean from India to Kenya (about 3000 km). Ten types of cuckoos perform polarized intra-continental migration in a way that they spend non-breeding seasons in tropical areas of the continent and then immigrate to dry and desert areas for egg laying.

About 52 old species and 3 new species are brood parasite. They lay their eggs in other birds' nests. These species are obligate brood parasites since this is the only way to their reproduction. Cuckoo eggs hatch earlier than their host's eggs. In most cases, a cuckoo chick throws the host's eggs or the host's chicks out of the nest. This is completely instinctive and the cuckoo chick has no time to learn it. A cuckoo chick makes the host provide a food suitable to its growth and beg for food again and again. The cuckoo chick announces its need for food by an open mouth because an open mouth to the mother is an indication for hunger.

Female cuckoos are skillful and expert in producing eggs similar to their host's eggs due to natural selection. Some birds recognize the eggs and throw them out though. Parasite cuckoos are divided into different categories and each category is expert in a special host. It is proved that the cuckoos in one category are genetically different from those in another category. Specialization in imitating the host's eggs gradually improves and evolves.

5.3 Details of COA

Figure 5.1 shows the flowchart of COA. Like other evolutionary algorithms, COA starts with an initial population (population of cuckoos). These cuckoos have got some eggs that will be laid in other species' nests. Some of these eggs that look like the host's eggs are more probable to be raised and turned into cuckoos. Other eggs are detected by the host and are demised. The rate of the raised eggs shows the suitability of the area. If there are more eggs to be survived in an area, there is more profits to that area. Thus the situation in which more eggs are survived will be a parameter for the cuckoos to be optimized.

Cuckoos search for the best area to maximize their eggs' life lengths. After hatching and turning into mature cuckoos, they form societies and communities. Each community has its habitat to live. The best habitat of all communities will be the next destination for cuckoos in other groups. All groups immigrate to the best current existing area. Each group will be the resident in an area near the best current existing area. An egg laying radius (*ELR*) will be calculated regarding the number of eggs each cuckoo lays and its distance from the current optimized area.

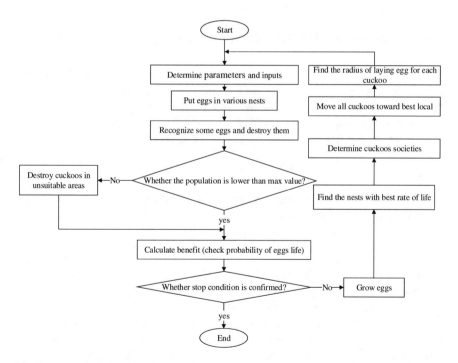

Fig. 5.1 Flowchart of the COA

Table 5.1 Characteristics of the COA

General algorithm	Cuckoo optimization algorithm
Decision variable	Cuckoo habitat
Solution	Habitat
Old solution	Old habitat
New solution	New habitat
Best solution	Habitat with best rate of life
Fitness function	Distance between best habitat and recent habitat
Initial solution	Random eggs for all cuckoos
Selection	–
Process of generating new solution	Emigration cuckoos toward best area

Afterwards, cuckoos start laying eggs randomly in the nests within their egg laying radii. This process continues until reaching the best place for egg laying (a zone with the most profit). This optimized zone is the place in which the maximum number of cuckoos gathers together. Table 5.1 lists the characteristics of the COA.

5.4 Cuckoos' Initial Residence Locations

It is necessary to form variables as an array so that an optimization problem can be solved. In GA and Particle Swarm Optimization, these arrays are identified by "chromosome" and "particles' positions", but in COA, this array is called "habitat".

In a one-dimensional *Nvar* optimization problem, habitat is a $1 \times Nvar$ array that shows the current position of cuckoos' life. It is defined as:

$$\text{Habitat} = [x_1, x_2, \ldots, x_{N\text{var}}] \tag{5.1}$$

The amount of profit or suitability rate for the current habitat can be obtained by profit function evaluation. Thus

$$\text{Porofit} = f_p(\text{habitat}) = f_p(x_1, x_2, \ldots, x_{N\text{var}}) \tag{5.2}$$

COA is an algorithm that maximizes the profit function. To use COA, the cost function should be multiplied by a minus sign so that the problem could be solved.

To start optimization, a habitat matrix sized $N_{\text{pop}} \times N_{\text{var}}$ is generated. Afterwards, a number of random eggs are specified for each habitat matrix. Each cuckoo lays 5–20 eggs in nature. These numbers are used as the maximum and minimum limits in the egg specification of each cuckoo in different iterations. Each real cuckoo lays eggs in a specific range. Thus, the maximum range of egg laying is the ELR. In an optimization problem, with the upper and lower limits of var_{hi} and var_{low}, each cuckoo has an ELR which is proportionate to the total number of eggs, current number of eggs, and the upper/lower limits of variables of the problem.

So ELR is defined as (Rajabioun 2011):

$$ELR = \alpha \times \frac{\text{Number of current cuckoos eggs}}{\text{Total number of eggs}} \times (var_{hi} - var_{low}) \qquad (5.3)$$

in which α is a variable, by which the maximum ELR is set.

5.5   Cuckoos' Egg Laying Approach

Each cuckoo randomly lays eggs in its host bird's nest within its ELR. Figure 5.2 shows the egg laying radius or the maximum range of egg laying.

After all of the cuckoos lay their eggs, some of the eggs which are less similar to the host's eggs are recognized and thrown out. Thus after each egg laying process, p % of all eggs (usually 10%) whose profit function value is less will be destroyed. The rest of chicks in the host's nest are fed and raised.

Another interesting point about cuckoo chicks is that only one egg has the opportunity to be raised in each nest. Because when cuckoo chicks hatch, they throw out the host's eggs. If the host's chicks hatch earlier, the cuckoo chick has eaten the largest amount of food (because its body is three times larger and it knocks other chicks over) and after several days the host's own chicks will die from hunger and only cuckoo chick will survive.

Fig. 5.2 Random egg laying in ELR (the black circle is the cuckoo's initial habitat with three eggs; and the white circles are the eggs at new positions)

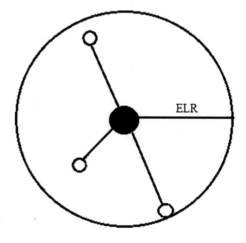

5.6 Cuckoos Immigration

When cuckoo chicks grow up and become mature, they live in the surrounding environment and in their communities for a while. But when the egg laying time is close, they immigrate to better habitats in which the chances for the survival of their eggs are higher. After forming cuckoo groups in various environments (search space of the problem), the group with the best position will be selected as the target group for other cuckoos for immigration.

It is difficult to recognize which group each cuckoo belongs when mature cuckoos live in several environment zones. To solve this problem, classification of cuckoos is done by K-means clustering (a number of K between 3 and 5 suffices).

The average profit of a group is calculated after all groups are formed to obtain the relative optimality of the living area of each group. Afterwards, the group with the highest value of average optimization will be selected as the target group and all others will immigrate toward this group. While immigrating toward the target point, cuckoos will not fly the whole way to the target place. They just pass a portion of the distance and they may even digress from the target too. This movement is shown in Fig. 5.3.

As shown in Fig. 5.3, each cuckoo only flies $\lambda\%$ of the entire distance toward the current ideal target and has a deviation of φ too. These two parameters help cuckoos search more space. λ is a random number between 0 and 1 and φ is a number from $\pi/6$ to $\pi/6$. When all cuckoos immigrate to the target point and their habitat points are determined, each cuckoo has a number of eggs. An ELR is determined for each cuckoo and then egg laying is started.

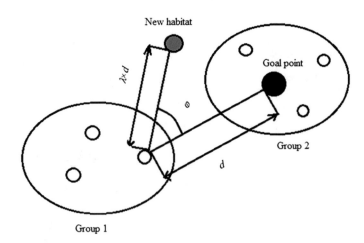

Fig. 5.3 Immigration of a sample cuckoo to the target habitat

5.7 Demising Cuckoos Laid in Inappropriate Positions

According to the fact that there is always a balance among the populations of birds in nature, a number N_{max} is used to control the maximum number of cuckoos that can live in a place. This balance is due to competing for limited foods, being hunted by predators, and finding improper nests for eggs.

5.8 Pseudo Code for COA

After several repetitions, all cuckoos will attain a point of optimization with the maximum similarity of their eggs to the host's eggs and with the maximum food sources. This location has the most total profit and the least chance for the eggs to be ruined. Convergence of more than 95% of all cuckoos toward one point will finalize the optimization process. The main steps of COA are shown in the following pseudo code:

Begin

    Define the number of habitats by generating X_i for i=1, 2, 3, ..., N

    Determine the upper and lower limits of the parameters of the optimization problem

(var_{hi}, var_{low})

    Consider the maximum number of cuckoos (N_{max})

    Specify the maximum and minimum numbers of eggs (E_{max}, E_{min})

    Specify the maximum number of iterations $(Iter_{max})$

    While (the stop criterion is not satisfied or $Iteration \leq Iter_{max}$)

        Define some cuckoos and assign some eggs to each cuckoo

        Calculate the radius of laying eggs for each cuckoo (ELR)

        Consider egg position or population (Xi_{new}) for each cuckoo in ELR

            If the population is lower than the maximum number of cuckoos $(Xi_{new}$

    $< N_{max})$

            Evaluate the fitness function of each egg (F_x)

        Else

            Destroy cuckoos in unsuitable areas

        End if

    End while

End

5.9   Capabilities of COA

The capabilities of COA in different fields can be summarized as follows:

1. Solving complicated non-linear optimization problems accurately;
2. Teaching artificial neuron networks, considering the fact that other approaches of optimization cannot provide an assured optimized solution due to the high number of parameters;
3. Using for easy and assured design of PID controllers for MIMO systems;
4. Finding the balance point in games quickly;
5. Optimizing the antenna design;
6. Optimizing the segmentation of pictures; and
7. Using for optimization problems that can be formulated as a target function.

5.10   Conclusion

The cuckoo optimization algorithm inspired by the life style of cuckoos was explained. The cuckoos' specific and unique feature in egg laying and raising is the basis of this algorithm. In COA, each cuckoo has a habitat in which eggs are laid. If the eggs survive, they are raised and become mature. Afterwards, they immigrate to the best habitat found for reproduction. The variety associated with cuckoos' movement toward the target habitat, provides more space for search. This algorithm is considered as a successful imitation of nature and is suitable for optimization problems in different fields.

References

Balochian, S., & Ebrahimi, E. (2013). Parameter optimization via cuckoo optimization algorithm of fuzzy controller for liquid level control. *Journal of Engineering*, 2013.

Kahramanli, H. (2012). A modified cuckoo optimization algorithm for engineering optimization. *International Journal of Future Computer and Communication, 1*(2), 199.

Khajeh, M., & Golzary, A. R. (2014). Synthesis of zinc oxide nanoparticles–chitosan for extraction of methyl orange from water samples: Cuckoo optimization algorithm–artificial neural network. *Spectrochimica Acta Part A, 131*, 189–194.

Khajeh, M., & Jahanbin, E. (2014). Application of cuckoo optimization algorithm–artificial neural network method of zinc oxide nanoparticles–chitosan for extraction of uranium from water samples. *Chemometrics and Intelligent Laboratory Systems, 135*, 70–75.

Mellal, M. A., & Williams, E. J. (2015a). Cuckoo optimization algorithm for unit production cost in multi-pass turning operations. *The International Journal of Advanced Manufacturing Technology, 76*(1–4), 647–656.

Mellal, M. A., & Williams, E. J. (2015b). Cuckoo optimization algorithm with penalty function for combined heat and power economic dispatch problem. *Energy, 93*, 1711–1718.

Mellal, M. A., & Williams, E. J. (2016). Parameter optimization of advanced machining processes using cuckoo optimization algorithm and hoopoe heuristic. *Journal of Intelligent Manufacturing, 27*(5), 927–942.

Moezi, S. A., Zakeri, E., Zare, A., & Nedaei, M. (2015). On the application of modified cuckoo optimization algorithm to the crack detection problem of cantilever Euler-Bernoulli beam. *Computers & Structures, 157,* 42–50.

Rabiee, M. and Sajedi, H. (2013). "Job scheduling in grid computing with cuckoo optimization algorithm." International Journal of Computer Applications, 62(16).

Rajabioun, R. (2011). Cuckoo optimization algorithm. *Elsevier, 11*(8), 5508–5518.

Singh, U., & Rattan, M. (2014). Design of linear and circular antenna arrays using cuckoo optimization algorithm. *Progress in Electromagnetics Research C, 46,* 1–11.

Shadkam, E., & Bijari, M. (2014). Evaluation the efficiency of cuckoo optimization algorithm. *International Journal on Computational Sciences and Applications (IJCSA), 4,* 39–47.

Shokri-Ghaleh, H., & Alfi, A. (2014). Optimal synchronization of teleoperation systems via cuckoo optimization algorithm. *Nonlinear Dynamics, 78*(4), 2359–2376.

Chapter 6
Teaching-Learning-Based Optimization (TLBO) Algorithm

Parisa Sarzaeim, Omid Bozorg-Haddad and Xuefeng Chu

Abstract This chapter is prepared to describe the Teaching-Learning-Based Optimization (TLBO) algorithm, a novel metaheuristic optimization method inspired by an educational classroom environment. It has an interesting exclusivity which may facilitate the solution process of optimization problems. In this chapter, a brief literature review of the TLBO algorithm is first presented. Then, the working process and two phases of TLBO (teacher phase and learner phase) are depicted. Eventually, a pseudocode of TLBO is presented.

6.1 Introduction

The Teaching-Learning-Based Optimization (TLBO) algorithm was first proposed by Rao et al. (2011). The elitist version of the TLBO algorithm, in which the worst individuals were replaced by elite individuals for the next generation, was developed by Rao and Patel (2012). The modified version of TLBO for multi-objective optimization problems was also developed by Rao and Patel (2013a, b). Despite the fact that the TLBO algorithm is a new metaheuristic optimization method, it has been applied to various engineering and science fields such as mechanical, civil,

P. Sarzaeim · O. Bozorg-Haddad (✉)
Department of Irrigation and Reclamation Engineering, Faculty of Agricultural Engineering and Technology, College of Agriculture and Natural Resources, University of Tehran, Karaj, Tehran 31587-77871, Iran
e-mail: OBHaddad@ut.ac.ir

P. Sarzaeim
e-mail: Parisa.Sarzaeim@ut.ac.ir

X. Chu
Department of Civil and Environmental Engineering, North Dakota State University, Dept 2470, Fargo, ND 58108-6050, USA
e-mail: Xuefeng.Chu@ndsu.edu

© Springer Nature Singapore Pte Ltd. 2018
O. Bozorg-Haddad (ed.), *Advanced Optimization by Nature-Inspired Algorithms*, Studies in Computational Intelligence 720, DOI 10.1007/978-981-10-5221-7_6

electrical, and environmental engineering. For example, Rao and Kalyankar (2012) used the TLBO algorithm for mechanical design optimization problems. Toğan (2012) optimized the design of planar steel frames by using the TLBO algorithm. Baghlani and Makiabadi (2013) used the TLBO algorithm to optimize the design of truss structures. García and Mena (2013) presented an optimum design of distributed generation by using a modified version of the TLBO algorithm. Roy (2013) and Roy et al. (2013) obtained an optimum solution to a hydrothermal scheduling problem for hydropower plants by the TLBO algorithm. Sultana and Roy (2014) applied the TLBO algorithm to minimize power loss and energy cost in power distribution systems. Bouchekara et al. (2014) applied the TLBO algorithm to solve the power flow problem. Ji et al. (2014) applied a modified TLBO algorithm to improve the forecast accuracy of water supply system operation. Bayram et al. (2015) used the TLBO algorithm to predict the concentrations of dissolved oxygen in surface water. Thus, the TLBO algorithm has a variety of applications because it is easy to use and convenient to be adapted for different problems.

6.2 Mapping a Classroom into the Teaching-Learning-Based Optimization Algorithm

All evolutionary and metaheuristic algorithms need some controlling parameters which vary in different problems. These controlling parameters are divided into two general groups: (1) general parameters such as population size, and number of generations or number of iterations, and (2) specific parameters which depend on the type of the algorithm. For instance, the Genetic Algorithm (GA) requires crossover rate and mutation rate. In the GA algorithms, proper selection of specific parameters has a significant effect on their performance, computation time, and modeling outputs. On the other hand, if the specific controlling parameters are selected improperly, the solution may be stuck in a local optimum or the improper parameters may lead to reduction of the solution quality. In such a situation, sensitivity analysis should be performed by user to identify the best measures of the specific parameters though it needs more time. Compared with other optimization algorithms, the TLBO algorithm does not require any of such specific parameters. It only requires general parameters such as population size and number of generations. It seems to be an interesting property which simplifies the application of the algorithm. Thus, TLBO algorithm is self-regulating.

The principle of the TLBO algorithm is inspired by the teacher–students relation in an educational classroom environment, the influence of teacher on learners or students, and the interactions of learners and their effects on each other. Teacher and learners are two main sectors of the algorithm which are named teacher phase and learner phase, respectively.

6.2.1   Teacher Phase

Imagine a classroom where there are two major groups: a teacher who teaches the class and some students who are learning. The responsibility of the teacher is to improve the knowledge level of the whole class to lead to better performance of students in exams. In the teacher phase, the teacher tries to transfer the knowledge to the learners. Thus, the teacher has a high knowledge level in the classroom and endeavors to raise the level of class. Suppose that there are n students ($j = 1, 2, \ldots, n$) in a classroom whose average grade in exam i is M_i and the best learner who achieves the best grade $X_{T,i}$ is supposed to be the teacher. The difference between the classroom average grade (M_i) and the best grade ($X_{T,i}$) can be computed by

$$\text{Diff}_i = r_i(X_{T,i} - T_F M_i) \tag{6.1}$$

where Diff_i = difference between the average grade and the best grade; r_i = random number in $[0, 1]$ in iteration i; $X_{T,i}$ = grade of the best learner (teacher) in iteration i; T_F = teacher factor which depends on teaching quality and is either 1 or 2; and M_i = average of learners' grades in iteration i.

T_F is also a random number which is given

$$T_F = \text{round}[1 + rand(0, 1)\{2 - 1\}] \tag{6.2}$$

Then, by using Diff_i, the new grade of student j in iteration i can be expressed as

$$X'_{j,i} = X_{j,i} + \text{Diff}_i \tag{6.3}$$

where $X'_{j,i}$ = new grade of student j in iteration i and $X_{j,i}$ = old grade of student j in iteration i. If $X'_{j,i}$ is better than $X_{j,i}$, $X'_{j,i}$ will go through to the learner phase. Otherwise, $X_{j,i}$ will go through to the learner phase.

6.2.2   Learner Phase

In a classroom, successful students try to help other students in order to increase their level of knowledge. In other words, the students help each other in some ways, for instance doing assignments in a group, to learn course materials better than the teacher teaches them only. Suppose that two students, students A and B, are selected randomly among the students in a classroom. The way that they help each other can be expressed as

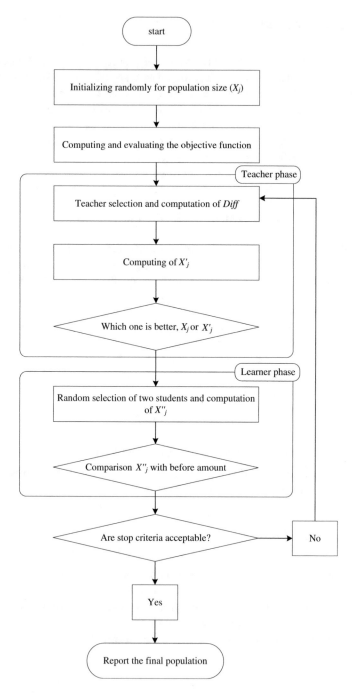

Fig. 6.1 Flowchart of the TLBO algorithm

$$X''_{A,i} = \begin{cases} X'_{A,i} + r_i(X'_{A,i} - X'_{B,i}) & \text{if} \quad X'_{A,i} > X'_{B,i} \\ X'_{A,i} + r_i(X'_{B,i} - X'_{A,i}) & \text{if} \quad X'_{B,i} > X'_{A,i} \end{cases} \tag{6.4}$$

If $X''_{A,i}$ is better than $X'_{A,i}$, $X''_{A,i}$ will go through the next iteration. Otherwise, $X'_{A,i}$ will go through the next iteration.

The steps of the working method of TLBO are summarized as follows:

(1) Initialize the grades $(X_{j,i})$ of n students (population size) randomly for iteration i.
(2) Evaluate the objective function for n students.
(3) Select the best objective function (teacher) and compute Diff_i by using Eq. (6.1).
(4) Compute $X'_{j,i}$ for n students and iteration i by using Eq. (6.3).
(5) Compare $X_{j,i}$ with $X'_{j,i}$. The better one goes to next step and the other one is removed.
(6) Select pairs of the students randomly and compare with each other. Then $X''_{j,i}$ is computed for every student by using Eq. (6.4).
(7) Compare $X''_{j,i}$ with $X'_{j,i}$. The better one goes to next step and the other one is removed.
(8) Evaluate the objective function for n students. Check if the stop criteria are satisfied. If yes, the best solution is achieved; otherwise, go back to Step (3).

The flowchart of the TLBO algorithm is shown in Fig. 6.1 to comprehend the working process of TLBO better and the algorithm's characteristics are shown in Table 6.1.

Table 6.1 Characteristics of the TLBO algorithm

General algorithm	Teaching-learning-based optimization
Decision variable	Grades of learners
Solution	Evaluating the grades of learners
Old solution	Old grades of learners
New solution	New grades of learners
Best solution	Best grade of learners
Fitness function	The grade is better than the one before
Initial solution	Randomly selected grades
Selection	–
Process of generating new solution	Comparison of the old and new grades of learners

6.3   Pseudo Code of the TLBO Algorithm

Begin

   Input $nPop$ = population size, It = number of iterations, and OB = objective function

   Initialize population members $X_{j,i}$ (learners) randomly for j=1, $nPop$

   Evaluate objective function (OB) for j=1, $nPop$

   While (until the stop criteria are satisfied) $i = 1, It$

      Calculation of population average M_i

      Selection of the best solution as teacher $X_{T,i}$

      Calculation of teacher factor $T_F = round[1 + rand(0,1)\{2-1\}]$

      Calculation of $Diff_i = r_i(X_{T,i} - T_F M_i)$

         For j=1, $nPop$

            Calculation of $X'_{j,i} = X_{j,i} + Diff_i$

            Evaluation of objective function (OB) of $X'_{j,i}$

            Comparison of $X_{j,i}$ with $X'_{j,i}$

               If $OB(X'_{j,i})$ is better than $OB(X_{j,i})$

               Then Replacement of $X_{j,i}$ by $X'_{j,i}$ ($X_{j,i(New)}$=

$X'_{j,i}$)

               End If

            Random selection of two different learners A and B $X_{j,i(New)}$

            Comparison of objective function of $X_{j\_A,i(New)}$ with

               $X_{j\_B,i(New)}$

               If $OB(X_{j\_A,i(New)})$ is better than $OB(X_{j\_B,i(New)})$

            Then

$X''_{j\_A,i} = X_{j\_A,i(New)} + r_i(X_{j\_A,i(New)} - X_{j\_B,i(New)})$

            Else

$X''_{j\_A,i} = X_{j\_A,i(New)} + r_i(X_{j\_B,i(New)} - X_{j\_A,i(New)})$

               End If

         End For

      Evaluation of the stop criteria

   End While

End

6.4 Conclusion

In this chapter, the teaching-learning-based optimization algorithm, a novel meta-heuristic optimization method, was described. First, a brief literature review of development and applications of the TLBO algorithm was presented. Then, fundamental details on the TLBO algorithm, including the two main phases (teacher phase and learner phase), all computational steps, and the optimization process were described. Finally, a pseudocode of TLBO was presented to demonstrate the implementation of this optimization approach. As aforementioned, the TLBO algorithm is a new user-friendly optimization algorithm that can be applied in different fields. As aforementioned, the TLBO algorithm does not require any specific parameters, and thus it is a valuable method that can achieve the final optimization solution, without the need to specify any parameters.

References

Baghlani, A., & Makiabadi, M. H. (2013). Teaching-learning-based optimization algorithm for shape and size optimization of truss structures with dynamic frequency constraints. *IJST, Transactions of Civil Engineering, 37,* 409–421.

Bayram, A., Uzlu, E., Kankal, M., & Dede, T. (2015). Modeling stream dissolved oxygen concentration using teaching-learning based optimization algorithm. *Environmental Earth Sciences, 73*(10), 6565–6576.

Bouchekara, H. R. E. H., Abido, M. A., & Boucherma, M. (2014). Optimal power flow using teaching-learning-based optimization technique. *Electric Power Systems Research, 114,* 49–59.

García, J. A. M., & Mena, A. J. G. (2013). Optimal distributed generation location and size using a modified teaching-learning based optimization algorithm. *Electrical Power and Energy Systems, 50,* 65–75.

Ji, G., Wang, J., Ge, Y., & Liu, H. (2014). Urban water demand forecasting by LS-SVM with tuning based on elitist teaching-learning-based optimization. In *Proceeding of 26th Chinese Control and Decision Conference*, Changsha, China, May 31–June 2.

Rao, R. V., Savsani, V. J., & Vakharia, D. P. (2011). Teaching-learning-based optimization: A novel method for constrained mechanical design optimization problems. *Computer-Aided Design, 43*(3), 303–315.

Rao, R. V., & Patel, V. (2012). An elitist teaching-learners-based optimization algorithm for solving complex constrained optimization problems. *International Journal of Industrial Computations, 3*(4), 535–560.

Rao, R. V., & Patel, V. (2013a). Multi-objective optimization of heat exchangers using a modified teaching-learning-based optimization algorithm. *Applied Mathematical Modelling, 37*(3), 1147–1162.

Rao, R. V., & Patel, V. (2013b). Multi-objective optimization of two stage thermoelectric cooler using a modified teaching-learning-based optimization algorithm. *Engineering Applications of Artificial Intelligence, 26*(1), 430–445.

Rao, R. V., & Kalyankar, V. D. (2012). Parameters optimization of machining process using a new optimization algorithm. *Materials and Manufacturing Processes, 27*(9), 978–985.

Roy, P. K. (2013). Teaching learning based optimization for short-term hydrothermal scheduling problem considering valve point effect. *Electrical Power and Energy systems, 53,* 10–19.

Roy, P. K., Sur, A., & Pradhan, D. K. (2013). Optimal short-term hydro-thermal scheduling using quasi-oppositional teaching learning based optimization. *Engineering Applications of Artificial Intelligence, 26*(10), 2516–2524.

Sultana, S., & Roy, P. K. (2014). Optimal capacitor placement in radial distribution systems using teaching learning based optimization. *Electrical Power and energy systems, 54,* 387–398.

Toğan, V. (2012). Design of planar steel frames using teaching-learning based optimization. *Engineering Structures, 34,* 225–232.

Chapter 7
Flower Pollination Algorithm (FPA)

Marzie Azad, Omid Bozorg-Haddad and Xuefeng Chu

Abstract This chapter is designed to describe the flower pollination algorithm (FPA) which is a new metaheuristic algorithm. First, the FPA applications in different problems are summarized. Then, the natural pollination process and the flower pollination algorithm are described. Finally, a pseudocode of the FPA is presented.

7.1 Introduction

The flower pollination algorithm (FPA) was proposed by Yang (2012) for global optimization. This new metaheuristic algorithm is inspired by the pollination phenomenon of flowing plants in nature. Yang et al. (2013) used the eagle strategy with FPA to balance exploration and exploitation. Sharawi et al. (2014) employed FPA in a wireless sensor network for efficient selection cluster heads and compared with the Low-Energy Adaptive Clustering Hierarchy (LEACH). The results indicated that FPA outperformed the LEACH. Sakib et al. (2014) used FPA and the bat

M. Azad · O. Bozorg-Haddad (✉)
Department of Irrigation and Reclamation Engineering, Faculty of Agricultural Engineering and Technology, College of Agriculture and Natural Resources, University of Tehran, 31587-77871 Karaj, Tehran, Iran
e-mail: OBHaddad@ut.ac.ir

M. Azad
e-mail: M.Azad.71@ut.ac.ir

X. Chu
Department of Civil and Environmental Engineering, North Dakota State University, Dept 2470, Fargo, ND 58108-6050, USA
e-mail: Xuefeng.Chu@ndsu.edu

© Springer Nature Singapore Pte Ltd. 2018
O. Bozorg-Haddad (ed.), *Advanced Optimization by Nature-Inspired Algorithms*,
Studies in Computational Intelligence 720, DOI 10.1007/978-981-10-5221-7_7

algorithm (BA) to solve continuous optimization problems. They tested and compared the two algorithms on the benchmark functions. Emary et al. (2014) applied FPA to a retinal vessel segmentation optimization problem. Platt (2014) used FPA in the calculation of dew point pressure in a system that exhibited double retrograde vaporization. El-henawy and Ismail (2014) combined FPA with the particle swarm optimization (PSO) algorithm to solve large integer programming problems and demonstrated that FPA was useful for betterment searching accuracy. Abdel-Raouf et al. (2014) formulated Sudoku puzzles as an optimization problem, and then employed a hybrid optimization method, flower pollination algorithm with the Chaotic Harmony Search (FPCHS) to obtain the optimal solutions. Yang et al. (2014) used a novel version of FPA to solve several multi-objective test functions. Trivedi et al. (2015) used FPA for optimization of relay coordination in a wide electrical network with the aim of increasing the selectivity and at the same time reducing the fault clearing time to improve reliability of the system. Bensouyad and Saidouni (2015) applied the discrete flower pollination algorithm (DFPA) for solving a graph coloring problem. Lukasik and Kowalski (2015) tested FPA for a number of continuous benchmark problems. Dubey et al. (2015) applied a modified flower pollination algorithm in the modern power systems to find out the solutions to economic dispatch problems solutions. They added a scaling factor to control the local pollination and compression of the exploitation stage to achieve the best solution. Bibiks et al. (2015) used DFPA in order to solve combinatorial optimization problems. Alam et al. (2015) applied the FPA technique for determining optimal parameters of a single diode and two diodes that were used to describe photovoltaic systems. In the design of a structural system, the optimal values of design variables cannot be obtained analytically and a structural engineering problem has different design constraints, so optimization is an important part of the structural design process. For this purpose Nigdeli et al. (2016) used FPA to solve structural engineering problems related to pin-jointed plane frames, truss systems, deflection minimization of I-beams, tubular columns, and cantilever beams. Nabil (2016) developed a Modified Flower Pollination Algorithm (MFPA) from the hybridization FPA with the Clonal Selection Algorithm (CSA) and performed tests on 23 optimization benchmark problems to investigate the efficiency of the new algorithm. Then, the results of MFPA were compared with those of Simulated Annealing (SA), Genetic Algorithm (GA), FPA, Bat Algorithm (BA), and Firefly Algorithm (FA). The results showed that the proposed MFPA was able to find more accurate solutions than FPA and the four other algorithms. Abdelaziz et al. (2016) applied FPA to drive the optimal sizing and allocations of the capacitors in different water distribution systems.

7.2 Flower Pollination Process

Pollination is a natural mechanism for the reproduction of flowering plants and is defined as transfer of pollen from one flower flag to the pistil stigma of the same flower or another flower of the same plant species. Pollen has both vegetative and reproductive cells. After sitting pollen on the pistil stigma, the vegetative cell multiplies and forms a pollen tube. A reproductive cell can be divided into two cells along its patch, reaching the ovary by a pollen tube. One of the reproductive cells is fertilized by an egg cell, forming a zygote. Thus, a new plant forms whit growth zygote. There are two types of pollination according to pollen transfer methods: (1) biotic pollination, and (2) abiotic pollination. For most flowering plants, biotic pollination is done by pollinators such as insects or animals. But abiotic pollination does not require the transfer of pollen by living organisms. Instead, it is done by water, wind, or gravity as pollinators. When pollens are delivered from one plant to another of the same type, such pollination is called cross-pollination and self-pollination occurs when pollen is delivered to the same flower or flowers of the same plant.

Almost 90% of flowering plants have biotic pollination in which the pollens are transferred by pollinators such as insects or animals. Pollination by insects is more relevant among plants. Flowers that pollinated with aim of insects attract insects by their colors, odors, and nectars. Generally, the size of flowers is consistent with the insects' bodies so that the insects can enter the flowers and their bodies are in contact with the pollen and pistil. Approximately 10% of flowering plants perform abiotic pollination that does not need pollinators and the pollens are transferred by wind or water. Most of the bush plants and trees do not need insects for pollination. Pollens of these plants are spread in air and transferred by wind. Although most of the pollens are being lost, some of them can be trapped by ripen stigma of female flowers. Wind pollination takes place in the plants that have female and male flowers and these flowers exist in separate trees. In such flowers, their pollens are released by shaking the stamens by wind. Numerous fine pollens of these flowers can travel long distances with wind. The stigma of a flower has feathery ramifications and is outside of the flower, which increases the chance of the stigma to trap pollen transferred by wind. For the flowers that are pollinated by wind, their petioles are absent or very small and they have no nectar. Pollinator insects are often associated with a specific flower type, which is defined as flower constancy. That is, the pollinators tend to sit on certain flower species. Therefore, flower constancy helps quantify the cost of searching for each of pollinators. For biotic pollination, the pollinators such as flies, birds, and bats can fly long distances. Thus, they can be considered as global pollination. Likewise, step-jump or flying of birds or bees can be described as levy flight.

7.3   Flower Pollination Algorithm

The biotic pollination, cross-pollination, abiotic pollination and self-pollination strategies are defined in domain optimization and embedded in the flower pollination algorithm. The pollination process includes a series of complex mechanisms in plant production strategies. A flower and its pollen gametes form a solution of the optimization problem. Flower constancy as a fitted solution is perceptible. In global pollination, the pollinators transfer pollen in long distances towards high fitting. On the other hand, local pollination within a limited area of a unique flower takes place under shading by wind or water. Global pollination occurs with a probability that is called switch probability. If this step is removed, local pollination replaces it. In the FPA algorithm four rules are followed: (1) live pollination and cross-pollination are considered as global pollination and the carriers or pollen pollinators move in a way that follows levy fight; (2) abiotic and self-pollination are considered as local pollination; (3) pollinators including insects can develop flower constancy. Flower constancy is production probability that is proportional to the similarity of two involved flowers; and (4) the interaction of global and local pollination can be controlled by switch probability. The first and third rules can be expressed as (Yang 2012):

$$x_i^{t+1} = x_i^t + \gamma \times L(\lambda) \times (g_* - x_i^t) \tag{7.1}$$

where x_i^t = pollen or solution vector at iteration t; g_* = the current best solution among all current generation solutions; γ = a scale factor for controlling step size; and L = strength of pollination, which is a step size related to the levy distribution. Levy flight is a bunch of random processes where the length of each jump follows the levy probability distribution function and has infinite variance. Following Yang (2012), L for a levy distribution is given by:

$$L \sim \frac{\lambda \times \Gamma(\lambda) \times \sin\frac{\pi\lambda}{2}}{\pi} \times \frac{1}{S^{1+\lambda}} \quad S \gg S_0 0, \tag{7.2}$$

where $\Gamma(\lambda)$ = standard gamma function.

For local pollination, the second and third rules are given by:

$$x_i^{t+1} = x_i^t + \varepsilon\left(x_j^t - x_k^t\right) \tag{7.3}$$

Table 7.1 Characteristics of the FPA

General algorithm	Flower pollination algorithm
Decision variable	Flowers or pollen gametes in each dimension
Solution	Flowers or pollen gametes
Old solution	Old flower or old pollen gamete
New solution	New flower or new pollen gamete
Best solution	Current best solution
Fitness function	–
Initial solution Selection	Random selection
Process of generating new solution	Flying and local random walk

where x_j^t and x_k^t = two pollens from different flowers of the same plant. Mathematically, if x_j^t and x_k^t come from the same species or are selected from the same population, this becomes a local random walk if ε has a uniform distribution in [0,1].

Table 7.1 lists the characteristics of the FPA and Fig. 7.1 shows the flowchart of the FPA.

7.4 User-Defined Parameters of the FPA

The size of the population of solutions (n), the scale factor for controlling step size (γ), the levy distribution parameter $[L(\lambda)]$, and the switch probability (P) are user-defined parameters in the FPA. Determining the optimal parameters of the FPA is a time-consuming work and needs to run the algorithm many times. It should be noted that the optimal parameters of the algorithm for one problem are different from those of other problems. Considering a mixture of parameters is an appropriate method for finding the suitable values of the algorithm parameters. The algorithm can be done for several times for one mixture of parameters, and the similar process can be repeated for other mixtures of parameters. Finally, the results for different sets of parameters can be compared and the best value can be determined. Yang (2012) suggested to start the modeling with $P = 0.5$ and $\lambda = 1.5$.

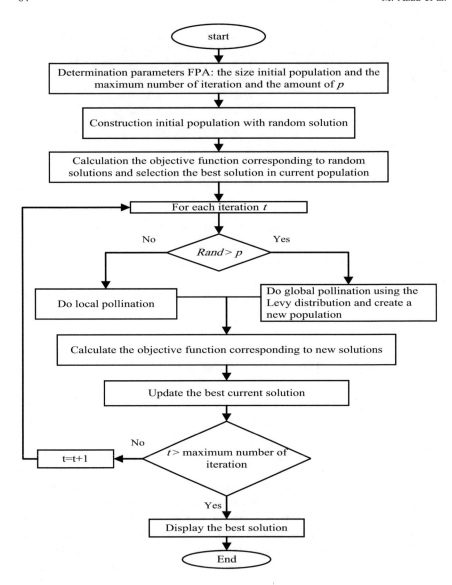

Fig. 7.1 Flowchart of the FPA

7.5   Pseudo Code of FPA

Begin

    Objective function: minimize or maximize $F(x)$, $x=(x_1, x_2, ..., x_d)$

    Initialize a population of n flowering species using random solutions

    Find the best solution g_* in the initial population

    Select a number in [0,1] with switch probability P

    While ($t <$ maximum iteration number)

        For $m= 1:n$

            If random $> P$ then

                Create a d-dimensional step vector L with the levy flight distribution

                Do global pollination with $x_i^{t+1} = x_i^t + \gamma \times L(\lambda) \times (g_* - x_i^t)$

            Else

                Create from a uniform distribution in [0,1]

                Select x_j^t and x_k^t from the population randomly

                Do local pollination with $x_i^{t+1} = x_i^t + (x_j^t - x_k^t)$

            End if

            Calculate the objective function for new solutions

            If the objective functions corresponding to new solutions are better, update new solutions

        End for

        Obtain current best solution g_*

    End while

End

7.6 Conclusion

This chapter described the flower pollination algorithm (FPA), which is based on the pollination phenomenon of flowing plants in nature. The chapter presented a summary of the FPA applications in different problems. Then, the natural process of pollination, and the flower pollination algorithm and a pseudocode of the FPA are presented.

References

Abdelaziz, A. Y., Ali, E. S., & Abd Elazim, S. M. (2016). Optimal sizing and locations of capacitors in radial distribution systems via flower pollination optimization algorithm and power loss index. *Engineering Science and Technology, 19*(1), 610–618.

Abdel-Raouf, O., & Abdel-Baset, M. (2014). A new hybrid flower pollination algorithm for solving constrained global optimization problems. *International Journal of Applied Operational Research-An Open Access Journal, 4*(2), 1–13.

Abdel-Raouf, O., El-Henawy, I., & Abdel-Baset, M. (2014). A novel hybrid flower pollination algorithm with chaotic harmony search for solving sudoku puzzles. *International Journal of Modern Education and Computer Science, 6*(3), 38.

Alam, D. F., Yousri, D. A., & Eteiba, M. B. (2015). Flower pollination algorithm based solar PV parameter estimation. *Energy Conversion and Management, 101,* 410–422.

Bekdaş, G., Nigdeli, S. M., & Yang, X. S. (2015). Sizing optimization of truss structures using flower pollination algorithm. *Applied Soft Computing, 37,* 322–331.

Bibiks, K., Li, J. P., & Hu, F. (2015). Discrete flower pollination algorithm for resource constrained project scheduling problem. *International Journal of Computer Science and Information Security, 13*(7), 8.

Dubey, H. M., Pandit, M., & Panigrahi, B. K. (2015). A biologically inspired modified flower pollination algorithm for solving economic dispatch problems in modern power systems. *Cognitive Computation, 7*(5), 594–608.

El-henawy, I., & Ismail, M. (2014). An improved chaotic flower pollination algorithm for solving large integer programming problems. *International Journal of Digital Content Technology and its Applications, 8*(3).

Emary, E., Zawbaa, H. M., Hassanien, A. E., Tolba, M. F., & Snášel, V. (2014). Retinal vessel segmentation based on flower pollination search algorithm. In *Proceedings of the Fifth International Conference on Innovations in Bio-Inspired Computing and Applications IBICA, 2014* (pp. 93–100). Springer International Publishing.

Łukasik, S., & Kowalski, P. A. (2015). Study of flower pollination algorithm for continuous optimization. In *Intelligent Systems, 2014* (pp. 451–459). Springer International Publishing.

Nabil, E. (2016). A modified flower pollination algorithm for global optimization. *Expert Systems with Applications, 57,* 192–203.

Nigdeli, S. M., Bekdaş, G., & Yang, X. S. (2016). Application of the flower pollination algorithm in structural engineering. In *Metaheuristics and optimization in civil engineering* (pp. 25–42). Springer International Publishing.

Platt, G. M. (2014). Computational experiments with flower pollination algorithm in the calculation of double retrograde dew points. *International Review of Chemical Engineering, 6* (2), 95–99.

Sakib, N., Kabir, M. W. U., Subbir, M., & Alam, S. (2014). A comparative study of flower pollination algorithm and bat algorithm on continuous optimization problems. *International Journal of Soft Computing and Engineering, 4*(2014), 13–19.

Sharawi, M., Emary, E., Saroit, I. A., & El-Mahdy, H. (2014). Flower pollination optimization algorithm for wireless sensor network lifetime global optimization. *International Journal of Soft Computing and Engineering, 4*(3), 54–59.

Trivedi, I. N., Purani, S. V., & Jangir, P. K. (2015). Optimized over-current relay coordination using Flower Pollination Algorithm. In *Advance Computing Conference (IACC), 2015 IEEE International* (pp. 72–77). IEEE.

Yang, X. S. (2012). Flower pollination algorithm for global optimization. In *International Conference on Unconventional Computing and Natural Computation* (pp. 240–249). Berlin: Springer.

Yang, X. S., Karamanoglu, M., & He, X. (2014). Flower pollination algorithm: A novel approach for multiobjective optimization. *Engineering Optimization, 46*(9), 1222–1237.

Yang, X. S., Deb, S., & He, X. (2013). Eagle strategy with flower algorithm. In *2013 International Conference on Advances in Computing, Communications and Informatics (ICACCI)* (pp. 1213–1217). IEEE.

Chapter 8
Krill Herd Algorithm (KHA)

Babak Zolghadr-Asli⑩, Omid Bozorg-Haddad and Xuefeng Chu

Abstract The krill herd algorithm (KHA) is a new metaheuristic search algorithm based on simulating the herding behavior of krill individuals using a Lagrangian model. This algorithm was developed by Gandomi and Alavi (2012) and the preliminary studies illustrated its potential in solving numerous complex engineering optimization problems. In this chapter, the natural process behind a standard KHA is described.

8.1 Introduction

In the past decades, metaheuristic optimization techniques have been widely employed in many fields to solve complex optimization problems, due to their advantages over the conventional optimization techniques. Generally, these algorithms have two main features: (1) intensification and (2) diversification (Gandomi et al. 2013c). The former denotes searching through the current candidates for the optimal solution, while the latter indicates expanding the searching horizon to ensure that the final result is a global optimum, instead of a local one. Each newly proposed algorithm attempts to improve these two main features, either by decreasing the distance of the reported solutions and the actual global optima or by reducing the solution searching time.

B. Zolghadr-Asli · O. Bozorg-Haddad (✉)
Department of Irrigation and Reclamation Engineering, Faculty of Agricultural Engineering and Technology, College of Agriculture and Natural Resources, University of Tehran, 3158777871 Karaj, Iran
e-mail: OBHaddad@ut.ac.ir

B. Zolghadr-Asli
e-mail: ZolghadrBabak@ut.ac.ir

X. Chu
Department of Civil and Environmental Engineering, North Dakota State University, Dept 2470, Fargo, ND 58108-6050, USA
e-mail: Xuefeng.Chu@ndsu.edu

O. Bozorg-Haddad (ed.), *Advanced Optimization by Nature-Inspired Algorithms*, Studies in Computational Intelligence 720, DOI 10.1007/978-981-10-5221-7_8

Although the basic principles of these algorithms are similar and contain an iterative mechanism, the iteration process differs in each algorithm. The main objective of such a process is to search through the decision space for arrays of decision variables that produce an optimum result. This process is usually inspired by the natural phenomena, and is intended to imitate a natural feature that has been evolved over millions of years (Gandomi and Alavi 2012). Consequently, there is no limitation to the source of inspiration for these bio-inspired algorithms, and they can imitate a vast domain of features, from the genetic evolution process of a species to the foraging mechanism of bacteria. Swarm intelligence, which is an imitation of an animal group's behavior, could serve as an inspiration source to develop such algorithms.

Many studies have focused on capturing the underlying mechanism that governs the development of formation grouping of various species of marine animals, including the Antarctic krill (Flierl et al. 1999). The krill herds are aggregations with no parallel orientation in both temporal and spatial scales (Brierley and Cox 2010). These creatures that can form large swamps are the source of inspiration for the krill herd algorithm (KHA). The herding of the krill individuals is a multi-objective process, including two main goals: (1) increasing krill density, and (2) reaching the food. Density-dependent attraction of krill (increasing density) and finding food (areas of high food concentration) are used as objectives, which finally cause krill to herd around the global optima. In this process, an individual krill moves toward the best solution when it searches for the highest density and food. The imaginary distance of krill individuals serve as objective functions, and minimizing them is the priority of the optimization process. Hence the closer the distance to the high density and food, the less the objective function (better) (Gandomi and Alavi 2012).

The engineering optimization problems, which mostly have a nonlinear decision space, are complicated, due to their numerous decision variables and complex constraints. Such conditions can be regarded as an advantage for the metaheuristic algorithms over the conventional optimization techniques. KHA is a new and novel metaheuristic search algorithm based on the herding behavior of krill individuals, using a Lagrangian model at its core. This algorithm was first introduced by Gandomi and Alavi (2012), and the preliminary studies have demonstrated its potential to outperform the existing algorithms for solving the complicated engineering problems (Gandomi et al. 2013a, b). Additionally, KHA was further validated for various engineering problems, including optimal design of civil structures (Gandomi et al. 2013a, b; Gandomi and Alavi 2016), power flow optimization (Mukherjee and Mukherjee 2015), and optimum operation of power plants (Mandal et al. 2014). Similarly, studies have illustrated that the power of the classical KHA tends towards global exploration (Bolaji et al. 2016). Some modifications have been made to the standard KHA, and the modified algorithms include: chaotic-particle swarm krill herd (CPKH) (Wang et al. 2013), fuzzy krill herd algorithm (FKH) (Fattahi et al. 2016), and discrete-based krill herd algorithm (DKH) (Bolaji et al. 2016). As a compatible and efficient algorithm, KHA can be a promising alternative for solving engineering optimization problems.

8.2 Krill Swarms' Herding Pattern

The basic core of a standard KHA is its krill herding simulator. The krill herd defuses after a hypothetical attack from a predator. This is the initial step in the standard KHA. Each krill after such an event has two priorities, which are decreasing its distance from both the food source and the highest density of the krill swarm. These imaginary distances are acting as the objective function, and minimizing these distances is considered as the goal of each krill individual. Consequently, the time-dependent position of an individual krill is governed by the motion induced by other krill individuals (N_i), foraging motion (F_i), and physical diffusion (D_i). As any efficient optimization algorithm should be compatible with arbitrary dimensions, since each arbitrary dimension is to represent a decision variable, the following Lagrangian model is generalized for an n-dimensional decision space (Gandomi and Alavi 2012):

$$\frac{dX_i}{dt} = N_i + F_i + D_i \tag{8.1}$$

in which X_i = location of the ith krill individual in the decision space.

Equation (8.1), which simulates the movement of each individual krill, implies that the movement of each krill is affected by three factors: (1) the behavior of the group, (2) the location of food, and (3) a random factor.

8.3 Motion Induced by the Krill Herd

Theoretically speaking, the krill herd has a tendency to move in a group. In other words, forming a high-density swarm is considered as an advantage for the krill community. Thus, krill individuals try to maintain a high density and move due to their mutual effects. The motion induced by the krill herd can be expressed as (Gandomi and Alavi 2012):

$$N_i^{new} = N^{max} \times \alpha_i + \omega_n \times N_i^{old} \tag{8.2}$$

in which N_i^{new} = motion of the ith krill individual induced by the krill herd at the current iteration; N^{max} = maximum induced speed (according to the measured values, it is around 0.01 m/s) (Hofmann et al. 2004); a_i = direction of motion of the ith krill individual induced by the krill swamp; w_n = inertia weight of the motion induced in the range [0,1]; and N_i^{old} = motion of the ith krill individual induced by the krill herd in the previous iteration. w_n is one of the model parameters and it acts as a weight for the previously calculated motion induced by the krill herd. A lower value of w_n decreases the influence of the N_i^{old}.

The direction of motion of the ith krill individual induced by the krill swamp (a_i), however, is influenced by both the nearby krill individuals (local effect) and the target swarm density (target effect), and it is given by Gandomi and Alavi (2012):

$$\alpha_i = \alpha_i^{\text{local}} + \alpha_i^{\text{target}} \tag{8.3}$$

$$\alpha_i^{\text{local}} = \sum_{j=1}^{NN} \widehat{K}_{(i,j)} \times \widehat{X}_{(i,j)} \tag{8.4}$$

$$\widehat{X}_{(i,j)} = \frac{X_j - X_i}{\|X_j - X_i\| + \varepsilon} \tag{8.5}$$

$$\widehat{K}_{(i,j)} = \frac{K_i - K_j}{K^{\text{worst}} - K^{\text{best}}} \tag{8.6}$$

in which $\widehat{X}_{(i,j)}$ = local effect induced by the jth neighboring krill individual for the ith krill individual; $\widehat{K}_{(i,j)}$ = target direction effect provided by the best krill individual; K_i and K_j = fitness values of the ith and jth krill individuals, respectively; K^{best} and K^{worst} = best and worst fitness values for krill individuals, respectively; and NN = number of neighboring krill individuals for the ith krill.

Equation (8.5) represents the unit vector that connects the ith krill to the jth krill, while Eq. (8.6) calculates the normalized fitness value, which plays the role of a weight for the unit vector in Eq. (8.5). In fact, each calculated $\widehat{X}_{(i,j)} \times \widehat{K}_{(i,j)}$ characterizes the effect of the jth neighboring krill. This influence can be (1) an attractive one $\left(\widehat{K}_{(i,j)} > 0\right)$, which indicates that both krill individuals are moving toward one another; (2) a repulsive one $\left(\widehat{K}_{(i,j)} < 0\right)$, which refers to a situation where both krill individuals are moving away from each other; and (3) a futile one $\left(\widehat{K}_{(i,j)} = 0\right)$, which suggests that both krill individuals are incurious toward one another. The summation of these weighted vectors shows the influence of the neighboring krill individuals on the motion induced by the ith krill.

To choose the number of neighboring krill individuals for any given krill, different strategies can be implemented. For instance, a neighboring ratio can be simply defined to find the number of the closest krill individuals. Reportedly, using the actual behavior of the krill individuals suggests that a sensing distance (d_s) is a proper value to determine the neighboring krill individuals (Fig. 8.1). The sensing distance for each krill individual in each iteration can be determined by Gandomi and Alavi (2012):

$$d_{(s,i)} = \frac{1}{5N} \sum_{j=1}^{N} \|X_i - X_j\| \tag{8.7}$$

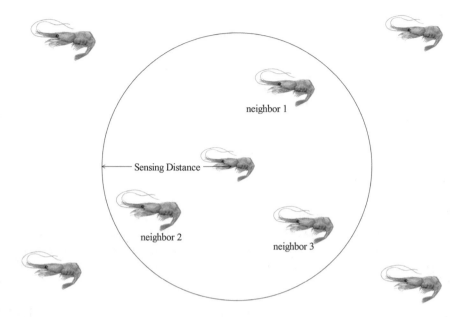

Fig. 8.1 Schematic representation of the sensing ambit around a krill individual

in which N = number of krill individuals. It should be noted that this is merely a suggestion and other techniques could be used, instead. In fact, the number of neighboring krill individuals (NN) is also one of the algorithm's parameters.

Equation (8.7) suggests that if the distance between two krill individuals is less than the calculated d_s, they should be considered as neighbors, and thus, they could influence the movement of one another in a direct manner. An analytical review of Eq. (8.7) also reveals that this formula has a tendency to calculate higher ratios of d_s for those krill individuals that are separated from the herd (placed in a low-density position). In this way there is a better chance that the neighbors of such a krill could make a greater impact on its motion so that it could decrease its distance from the herd.

The known target vector of each krill swarm has the lowest fitness of an individual krill in that herd. The effect of the individual krill with the best fitness on the ith individual krill is taken into account using Eq. (8.8). Such an effect can increase the chance of finding the global optima (Gandomi and Alavi 2012).

$$\alpha_i^{\text{target}} = C^{\text{best}} \times \widehat{K}_{(i,\text{best})} \times \widehat{X}_{(i,\text{best})} \qquad (8.8)$$

in which α_i^{target} = effective coefficient of the krill individuals with the best fitness to the ith krill individual. α_i^{target} leads the solution to the probable location of global optima and hence it should be more effective than other krill individuals such as neighboring krill individuals. Herein, C^{best} in Eq. (8.8) is defined as (Gandomi and Alavi 2012):

$$C^{\text{best}} = 2\left(\text{rand} + \frac{I}{I_{\max}}\right) \tag{8.9}$$

where rand = a random value in the range of [0,1] and it has a uniform distribution; I = current iteration number; and $I_{\max}$ = maximum number of iterations. Equation (8.9) suggests that the effect of the target krill is enhanced in each iteration.

8.4 Foraging Motion

The foraging motion, which is centered around the krill herd's tendency to find nutrition, has two terms in its structure: (1) the location of food, and (2) the previous experiences about the food location encountered by each individual krill. The above mechanism can be formulated for each individual krill as follows (Gandomi and Alavi 2012):

$$F_i = V_f \times \beta_i + \omega_f \times F_i^{\text{old}} \tag{8.10}$$

where

$$\beta_i = \beta_i^{\text{food}} + \beta_i^{\text{best}} \tag{8.11}$$

in which V_f = foraging speed (according to the measured values, it is 0.02 m/s) (Price 1989); w_f = inertia for the foraging motion in the range of [0,1]; β_i^{food} = food attraction parameter; and β_i^{best} = effect value of the best fitness of the ith krill.

Equations (8.10) and (8.11) encourage each individual krill to decrease its distance from the location of food, which is the probable location of the global optima in the decision space. However, the location of food and its attraction are, in fact, the result of the optimization process. In each iteration, the location of food and its attraction for the krill herd can only be estimated. While the estimation process can also influence the KHA efficacy, the following formula can be adapted to approximate the virtual center of the food concentration, using the fitness distribution of krill individuals. This approach, inspired by the "center of the mass" notion, can be expressed as (Gandomi and Alavi 2012):

$$X^{\text{food}} = \frac{\sum\limits_{i=1}^{N} \frac{1}{K_i} \times X_i}{\sum\limits_{i=1}^{N} \frac{1}{K_i}} \tag{8.12}$$

in which X^{food} = estimated location of the food.

Thus, the attraction of the food is given by

$$\beta_i^{\text{food}} = C^{\text{food}} \times \widehat{K}_{(i,\text{food})} \times \widehat{X}_{(i,\text{food})} \qquad (8.13)$$

in which C^{food} = food concentration. To ensure that the food attraction is decreasing for the krill herd during the iteration procedure, this food concentration term is added to the food attraction calculation process, and it is given by Gandomi and Alavi (2012):

$$C^{\text{food}} = 2\left(1 - \frac{I}{I_{\text{max}}}\right) \qquad (8.14)$$

The main reason behind the food attraction is to ensure that the krill swarm finds the global optima. As a result, when the krill herd is randomly spread through the decision space, this motion can help the herd to gather around the plausible location of the food (global optima). However, as the searching process advances in each iteration, the herd must be able to spread in a limited space, to locate the best solution. Thus, as shown in the formula, this motion decreases with time. This can be considered as an efficient global optimization strategy that helps improve the efficiency of KHA.

Each individual krill is also moving due to its visited memory of the previously spotted locations of food. The effect of the best fitness of the ith krill individual can also be expressed as (Gandomi and Alavi 2012):

$$\beta_i^{\text{best}} = \widehat{K}_{(i,\text{best})} \times \widehat{X}_{(i,\text{best})} \qquad (8.15)$$

in which $K_{(i,best)}$ = best previously encountered position of the ith krill individual.

8.5 Physical Diffusion

The two mechanisms behind inducing motion to each individual krill (motion induced by the krill herd and foraging motion) are to ensure that after the initial separation of the krill herd throughout the decision space, the herd gathers around what is considered to be the global optima. Yet, to ensure that the decision space is inspected thoroughly by the krill herd, a random process is required to spread an enough number of krill individuals in the decision space, in a random-based manner. If the random process is too strong, the herd will not gather around a center location; yet, lack of such a mechanism could interrupt a proper search throughout the decision space. The physical diffusion term is introduced in the KHA as a random process. This motion can be expressed in terms of maximum diffusion speed, a random direction vector, and a mathematical mechanism to ensure the decreasing effects of this term as searching for the global optimal solution continues. Thus, the physical diffusion term can be formulated as follows (Gandomi and Alavi 2012):

$$D_i = D^{\text{max}} \times \left(1 - \frac{I}{I_{\text{max}}}\right) \times \delta \tag{8.16}$$

in which D^{max} = maximum diffusion speed, which has the range of [0.002, 0.010] (m/s) (Gandomi and Alavi 2012); and δ = random directional vector and its arrays are random values between -1 and 1. A random selection can also be employed to determine the value of D^{max}. The physical diffusion motion introduced in Eq. (8.16) works on the basis of a geometrical annealing schedule, and the random speed linearly decreases with time.

8.6 Motion Process of the KHA

The above three mechanisms allow one to calculate the direction and speed of relocation for each individual krill at any given iteration. In other words, the defined motions frequently change the position of a krill individual toward the position that is expected to be the best one. The motion induced by other krill individuals and the foraging motion are working in parallel, which resultantly makes the KHA a potentially powerful algorithm for solving complex optimization problems. The KHA formulation suggests that if any of K_j, K^{best}, K^{food}, and K_i^{food} can illustrate a better performance than the ith krill individual, they can have an attractive effect, which can inspire this krill to move toward any of these locations, in the hope that such an action would improve its fitness value. Such a mechanism can also have a negative effect. The K_j, K^{best}, K^{food}, and K_i^{food} can repulse the ith krill individual, causing it to move away from the aforementioned locations. Additionally, the physical diffusion can spread the krill herd throughout the decision space for a comprehensive search of the plausible arrays of decision variables. After calculating the motion for every krill in the herd, the position vector of the ith krill individual during the time interval from t to $t + Dt$ is given by

$$X_i(t + \Delta t) = X_i(t) + \Delta t \times \frac{dX_i}{dt} \tag{8.17}$$

It should be noted that Δt is one of the most important model parameters since it works as a scale factor for the speed vector. Thus, it should be carefully set for the optimization problem. Suggestively, Δt, which completely depends on the search space, can be estimated by Gandomi and Alavi (2012):

$$\Delta t = C_t \times \sum_{j=1}^{NV} \left(UB_j - LB_j\right) \tag{8.18}$$

in which NV = number of variables; and LB_j and UB_j = lower and upper bounds of the jth variable, respectively. It is empirically found that C_t is a constant within

(0,2]. It is also obvious that the low values of C_t let the krill individual search the space in a slower, yet more careful pace. One should bear in mind that, this parameter is the most important parameter of the model, and thus, needs to be carefully adapted to each optimization problem.

Finally, it should be pointed out that although the above-mentioned mechanisms are the core concept of a standard KHA, this algorithm is compatible to implement a few external searching operators, including but not limited to genetic operators

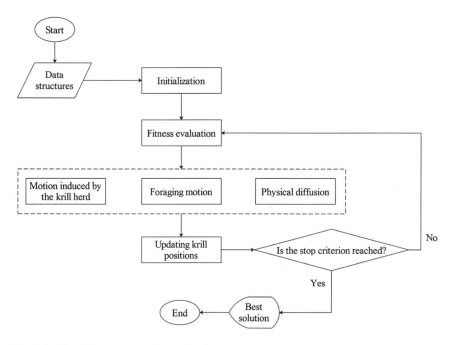

Fig. 8.2 Simplified flowchart of the KHA

Table 8.1 The characteristics of the KHA

General algorithm	Krill herd algorithm
Decision variable	Krill individual's position in each dimension
Solution	Krill individual's position
Old solution	Old position of the krill individual
New solution	New position of the krill individual
Best solution	Any krill with the best fitness
Fitness function	Distance between krill individual and food and the densest location in the herd
Initial solution	Randomly
Selection	–
Process of generating new solution	Motion induced by krill herd, foraging activity, and physical diffusion

such as crossover and mutation. While this capability surely enhances the performance of the standard KHA, their presence is not obligatory (Gandomi and Alavi 2012). A basic representation of the KHA is shown in Fig. 8.2. Additionally, Table 8.1 summarizes the characteristics of the standard KHA.

8.7   Pseudo Code of KHA

Begin

   Define population size (N) and maximum iteration number (I_{max})

   Set the iteration counter $I=1$

   Initialize the population by generating X_i for i = 1, 2, 3..., N

   Set the inertia weight of the motion induced (ω_n) and foraging motion (ω_f)

   Define Δt

   Evaluate each krill individual according to its position

   While (the stop criterion is not satisfied or $I < I_{max}$)

      For i = 1: N

         Perform the following motion calculation:

            Movement induced by other krill individuals

            Foraging activity

            Physical diffusion

         Update the krill individual position in the search space

         Evaluate each krill individual according to its position
      End for i

      Sort the population/krill from best to worst and find the current best

   End while

   Post-processing the results and visualization.

End

8.8 Conclusion

This chapter described the krill herd algorithm (KHA), which is a novel, yet relatively newly introduced metaheuristic optimization algorithm. After a brief review of the vast applications of KHA, including complex engineering optimization problems, the standard KHA and its mechanism were described. In the final section, a pseudo code of the standard KHA was also presented.

References

Bolaji, A. L. A., Al-Betar, M. A., Awadallah, M. A., Khader, A. T., & Abualigah, L. M. (2016). A comprehensive review: Krill Herd algorithm (KH) and its applications. *Applied Soft Computing, 49*, 437–446.

Brierley, A. S., & Cox, M. J. (2010). Shapes of krill swarms and fish schools emerge as aggregation members avoid predators and access oxygen. *Current Biology, 20*(19), 1758–1762.

Fattahi, E., Bidar, M., & Kanan, H. R. (2016). Fuzzy krill herd (FKH): An improved optimization algorithm. *Intelligent Data Analysis, 20*(1), 153–165.

Flierl, G., Grünbaum, D., Levins, S., & Olson, D. (1999). From individuals to aggregations: The interplay between behavior and physics. *Journal of Theoretical Biology, 196*(4), 397–454.

Gandomi, A. H., & Alavi, A. H. (2012). Krill herd: A new bio-inspired optimization algorithm. *Communications in Nonlinear Science and Numerical Simulation, 17*(12), 4831–4845.

Gandomi, A. H., & Alavi, A. H. (2016). An introduction of krill herd algorithm for engineering optimization. *Journal of Civil Engineering and Management, 22*(3), 302–310.

Gandomi, A. H., Alavi, A. H., & Talatahari, S. (2013a). Structural optimization using krill herd algorithm. Chapter 15 in *swarm intelligence and bio-inspired computation: Theory and applications*. London, UK: Elsevier Publication.

Gandomi, A. H., Talatahari, S., Tadbiri, F., & Alavi, A. H. (2013b). Krill herd algorithm for optimum design of truss structures. *International Journal of Bio-Inspired Computation, 5*(5), 281–288.

Gandomi, A. H., Yang, X. S., & Alavi, A. H. (2013c). Cuckoo search algorithm: A metaheuristic approach to solve structural optimization problems. *Engineering with Computers, 29*(1), 17–35.

Hofmann, E. E., Haskell, A. E., Klinck, J. M., & Lascara, C. M. (2004). Lagrangian modelling studies of Antarctic krill (*Euphausia superba*) swarm formation. *ICES Journal of Marine Science, 61*(4), 617–631.

Mandal, B., Roy, P. K., & Mandal, S. (2014). Economic load dispatch using krill herd algorithm. *International Journal of Electrical Power & Energy Systems, 57*, 1–10.

Mukherjee, A., & Mukherjee, V. (2015). Solution of optimal power flow using chaotic krill herd algorithm. *Chaos, Solitons & Fractals, 78*, 10–21.

Price, H. J. (1989). Swimming behavior of krill in response to algal patches: A mesocosm study. *Limnology and Oceanography, 34*(4), 649–659.

Wang, G. G., Gandomi, A. H., & Alavi, A. H. (2013). A chaotic particle-swarm krill herd algorithm for global numerical optimization. *Kybernetes, 42*(6), 962–978.

Chapter 9
Grey Wolf Optimization (GWO) Algorithm

Hossein Rezaei, Omid Bozorg-Haddad and Xuefeng Chu

Abstract This chapter describes the grey wolf optimization (GWO) algorithm as one of the new meta-heuristic algorithms. First, a brief literature review is presented and then the natural process of the GWO algorithm is described. Also, the optimization process and a pseudo code of the GWO algorithm are presented in this chapter.

9.1 Introduction

Grey wolf optimization (GWO) is one of the new meta-heuristic optimization algorithms, which was introduced by Mirjalili et al. (2014). Gholizadeh (2015) developed the GWO algorithm to solve an optimization problem of double-layer grids considering the nonlinear behavior. The results illustrated that GWO had a better performance than other algorithms in finding the optimal design of nonlinear double-layer grids. Mirjalili (2015) used the GWO algorithm to learn multi-layer perceptron (MLP) for the first time. In the study, the results of GWO were compared with those from particle swarm optimization (PSO), genetic algorithm (GA), ant colony optimization (ACO), and evolution strategy (EA), and indicated the higher performance of GWO. Saremi et al. (2015) coupled GWO with the evolutionary population dynamic (EPD) to improve the performance of the basic GWO

H. Rezaei · O. Bozorg-Haddad (✉)
Department of Irrigation and Reclamation Engineering, Faculty of Agricultural Engineering and Technology, College of Agriculture and Natural Resources, University of Tehran, 31587-77871 Karaj, Tehran, Iran
e-mail: OBHaddad@ut.ac.ir

H. Rezaei
e-mail: HosseinRezaie18@ut.ac.ir

X. Chu
Department of Civil and Environmental Engineering, North Dakota State University, Dept 2470, Fargo, ND 58108-6050, USA
e-mail: Xuefeng.Chu@ndsu.edu

© Springer Nature Singapore Pte Ltd. 2018
O. Bozorg-Haddad (ed.), *Advanced Optimization by Nature-Inspired Algorithms*,
Studies in Computational Intelligence 720, DOI 10.1007/978-981-10-5221-7_9

algorithm by removing weak individuals from the society. Comparison with the basic GWO illustrated that the proposed algorithm had a better performance in conversion rate and exploration, and also avoided trapping into local optima. Sulaiman et al. (2015) used GWO to solve an optimal reactive power dispatch (ORPD) problem and compared with swarm intelligence (SI), evolutionary computation (EC), PSO, harmony search algorithm (HAS), gravity search algorithm (GSA), invasive weed optimization, and modified imperialist competitive algorithm with invasive weed optimization (MICA-IWO). The results demonstrated that GWO had more desirable optimal solution than others.

9.2   Natural Process of the GWO Algorithm

GWO is inspired by social hierarchy and the intelligent hunting method of grey wolves. Usually, grey wolves are at the top of the food chain in their life areas. Grey wolves mostly live in a pack of 5–12 individuals. In particular, in grey wolves' life there is a strict social hierarchy. As shown in Fig. 9.1, the leaders of a pack of grey wolves (alpha) are a male and female wolves that often are responsible for making decisions for their pack such as sleep place, hunting, and wake-up time. Mostly, other individuals of the pack must obey the decision made by alpha. However, some democratic behaviors in the social hierarchy of grey wolves can be observed (alpha may follow other individuals of the pack). In gatherings, individuals confirm the alpha's decision by holding their tails down. It is also interesting to know that it

Fig. 9.1 Social hierarchy of grey wolves

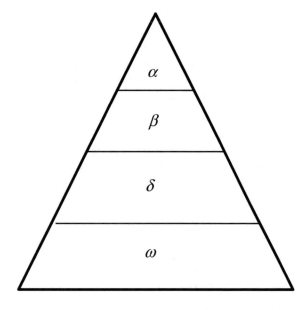

is not necessary for the alpha to be the strongest ones in the pack. Managing the pack is the main role of the alpha. In a pack of grey wolves, discipline and organization are the most important. The level next to alpha in the social hierarchy of grey wolves is beta and the role of beta is to help alpha in making decisions. Beta can be either male or female wolves and beta can be the best candidate of substitution for alpha when one of them becomes old or dies. The beta must respect alpha, but he/she can command other individuals. Beta is the consultant of alpha and responsible for disciplining the pack. The beta reinforces the orders of alpha and gives alpha the feedbacks. The weakest level in a pack of grey wolves is omega that plays a role of scapegoat. The wolves at the level of omega have to obey other individuals' orders and they are the last wolves that are allowed to eat food. Omega seems to be the least important individuals in the pack, but without omega, internal fight and other problems can be observed. This can be attributed to the omega's venting role of violence and frustration of other wolves, which helps satisfy other individuals and maintain the dominant structure of grey wolves. Sometimes, omega plays the role of babysitter in the pack. The remaining wolves, other than alpha, beta, and omega, are called subordinate (delta). The wolves at the level of delta obey the alpha and beta wolves and dominate the omega wolves. They act as scouts, sentinels, elders, hunters, and caretakers in the pack. Scouts are responsible for looking after boundaries and territory and also they should alarm the pack in facing to danger. Sentinels are in charge of security establishment. Elders are the experienced wolves that are candidates for alpha and beta. Hunters help alpha and beta in hunting and preparing food for the pack, while caretakers should look after the weak, ill, and wounded wolves.

In addition to the social hierarchy in a pack of grey wolves, group hunting is one of the interesting social behaviors of grey wolves too. According to Muro et al. (2011) grey wolves' hunting includes the following three main parts:

(1) Tracking, chasing, and approaching the prey.
(2) Pursuing, encircling, and harassing the prey till it stops moving.
(3) Attacking the prey.

These two social behaviors of grey wolves' pack (social hierarchy and hunting technique) are modeled in the GWO algorithm.

9.3 Mathematical Model of the GWO Algorithm

In this section, mathematical modeling of the social hierarchy of grey wolves, and their hunting technique (tracking, encircling, and attacking prey) in the GWO algorithm is detailed.

9.3.1 Social Hierarchy

In order to mathematically model the social hierarchy of grey wolves in the GWO algorithm, the best solution is considered as alpha (α). Therefore, the second and third best solutions are respectively considered as beta (β) and delta (δ), and other solution is assumed to be omega (ω). In the GWO algorithm, hunting (optimization) is guided by α, β, and δ, and ω wolves follow them.

9.3.2 Encircling the Prey

As aforementioned, grey wolves in the process of hunting, encircle the prey. The grey wolves' encircling behavior to hunt for a prey can be expressed as (Mirjalili et al. 2014):

$$\overrightarrow{D} = \left| \overrightarrow{C}.\overrightarrow{X}_p(t) - \overline{X}(t) \right| \tag{9.1}$$

$$\overrightarrow{X}(t+1) = \overrightarrow{X}_p(t) - \overrightarrow{A}.\overrightarrow{D} \tag{9.2}$$

where $t =$ iteration number; $\overrightarrow{A}$ and $\overrightarrow{C} =$ coefficient vectors; $\overrightarrow{X}_P =$ vector of the prey's positions; $\overrightarrow{X} =$ vector of the grey wolf's positions; and $\overrightarrow{D} =$ calculated vector used to specify a new position of the grey wolf. $\overrightarrow{A}$ and $\overrightarrow{C}$ can be calculated by Mirjalili et al. (2014):

$$\overrightarrow{A} = 2\overrightarrow{a}.\overrightarrow{r_1} - \overrightarrow{a} \tag{9.3}$$

$$\overrightarrow{C} = 2.\overrightarrow{r_2} \tag{9.4}$$

where $\overrightarrow{a} =$ vector set to decrease linearly from 2 to 0 over the iterations; and $\overrightarrow{r_1}$ and $\overrightarrow{r}_2 =$ random vectors in [0,1]. As shown in Fig. 9.2, a grey wolf at (x, y) can change its position based on the position of prey at (x', y'). Different places to the best agent can be achieved with respect to the current position by regulating the $\overrightarrow{A}$ and $\overrightarrow{C}$. For instance, by setting $\overrightarrow{A} = [1, 0]$ and $\overrightarrow{C} = [1, 1]$, the position of the grey wolf is updated to $(x' - x, y')$.

Note that the random $\overrightarrow{r_1}$ and $\overrightarrow{r_2}$ vectors let the grey wolf select any positions/nodes in Fig. 9.2. Therefore, a grey wolf can be placed in each random position around the prey that is calculated by using Eqs. (9.1) and (9.2). Following the same way, in an n-dimensional decision space grey wolves can move to any nodes of a hypercube around the best solution (position of the prey). They can distinguish the position of the prey from others and encircle it. Usually, hunting operation is guided by α, and β and δ provide support for α. In a decision space of an optimization problem we do not have any idea about the optimum solution.

Fig. 9.2 Attacking toward
prey versus searching for prey

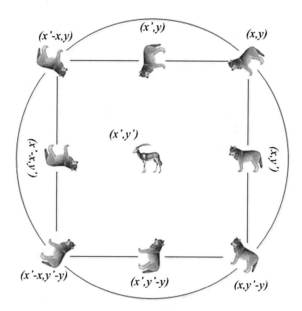

Thus, in order to simulate the hunting behavior of grey wolves, we assume that α (best candidate for the solution), β, and δ have more knowledge about the potential position of the prey. Therefore, the algorithm saves three best solutions achieved so far and forces others (i.e., omega wolves) to update their positions to achieve the best place in the decision space. In the optimization algorithm, such a hunting behavior can be modeled by Mirjalili et al. (2014):

$$\vec{D}_\alpha = \left| \vec{C}_1.\vec{X}_a - \vec{X} \right|, \ \vec{D}_\beta = \left| \vec{C}_2.\vec{X}_\beta - \vec{X} \right|, \ \vec{D}_\delta = \left| \vec{C}_3.\vec{X}_\delta - \vec{X} \right| \qquad (9.5)$$

$$\vec{X}_1 = \vec{X}_\alpha - A_1.(\vec{D}_\alpha), \ \vec{X}_2 = \vec{X}_\beta - A_2.(\vec{D}_\beta), \ \vec{X}_3 = \vec{X}_\delta - A_2.(\vec{D}_\delta) \qquad (9.6)$$

$$\vec{X}_1 = \vec{X}_\alpha - A_1.(\vec{D}_\alpha) \qquad (9.7)$$

Figure 9.3 shows how the search agent updates the positions of α, β, and δ in a 2D search space. As shown in Fig. 9.3, the final position (solution) is inside a circle that is specified based on the positions of α, β, and δ in the decision space. In other words, α, β, and δ estimate the positions of prey and other wolves and then update their new positions, randomly around the prey.

9.3.3   Attacking the Prey

As aforementioned, grey wolves finish the hunting process by attacking the prey until it stops moving. In order to model the attacking process, the value of $\vec{a}$ can be

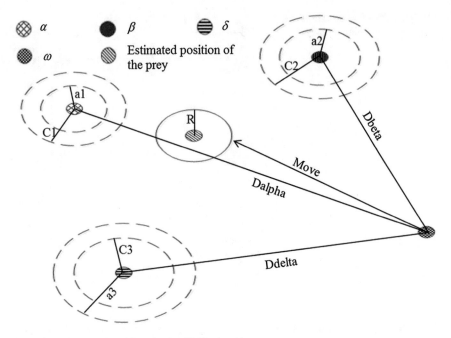

Fig. 9.3 Updating of positions in the GWO algorithm

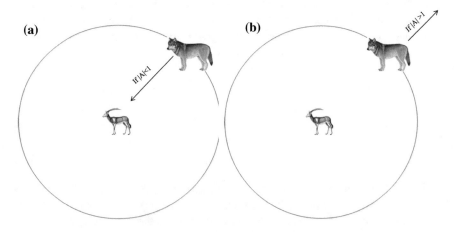

Fig. 9.4 Attacking toward prey and searching for prey

decreased in different iterations. Note that as $\vec{a}$ decreases the fluctuation rate of $\vec{A}$ decreases too. In other words, $\vec{A}$ is a random value in the range of $[-2a, 2a]$ where a decreases from 2 to 0 over iterations. When the random value of $\vec{A}$ is being in the range of $[-1, 1]$. The next position of a wolf can be between the current position and the prey position. As illustrated in Fig. 9.4, when $|A| < 1$ grey wolves will attack the prey.

By using the operators provided so far, the GWO algorithm lets the search agent to update its position based on the positions of α, β, and δ(move toward the prey). It is true that the encircling process provided as an operator in the GWO algorithm limits the solutions around local optima, but GWO also has many other operators to discover new solutions.

9.3.4   Searching for the Prey (Exploration)

Grey wolves often search for the prey according to the positions of α, β, and δ. They diverge from each other to explore the position of prey and then converge to attack the prey. In order to mathematically model the divergence of grey wolves, $\overrightarrow{A}$ can be utilized. $\overrightarrow{A}$ is a random vector that is greater than 1 or less than -1 to force the search agent to diverge from the prey, which emphasizes the global search in GWO. Figure 9.4 illustrates that when $\left|\overrightarrow{A}\right| > 1$, the grey wolf is forced to move away from the prey (local optimum) to search for better solutions in the decision space.

The GWO algorithm has another component $\left(\overrightarrow{C}\right)$ that assists the algorithm to discover new solutions. As shown in Eq. (9.4), the elements of vector $\overrightarrow{C}$ are within the range of $[0, 2]$. This component provides random weights for the prey to randomly emphasize $(C > 1)$ or deemphasize $(C < 1)$ the impact of the prey in defining the distance in Eq. (9.1). This component helps the GWO algorithm to behave more randomly and in favor of exploration, and keep the search agent away from local optima during the optimization process. Note that unlike A, C decreases nonlinearly. C is required in the GWO algorithm because not only in the initial iteration but also in the final iteration, it provides a global search in the decision space. This component is very useful in avoidance of local optima, especially in the

Table 9.1   Characteristics of the GWO algorithm

General algorithm	Grey wolf optimization algorithm
Decision variable	Grey wolf
Solution	Position of grey wolf
Old solution	Old position of grey wolf
New solution	New position of grey wolf
Best solution	Position of prey
Fitness function	Distance between grey wolf and prey
Initial solution	Initial random position of grey wolf
Selection	–
Process of generating new solution	Hunting operators (encircling and attacking prey)

final iteration. The C vector can be used as a hedge of approaching the prey in nature. Generally, the hedge can be seen in a nature hunting process of grey wolves. This hunting technique prevents grey wolves from quickly approaching the prey (this is truly what C does in the optimization process of the GWO algorithm). Table 9.1 presents the characteristics of the GWO algorithm.

9.4 Optimization Process in GWO Algorithm

The optimization process of GWO starts with creating random population of grey wolves (candidate solutions). Over the iterations, α, β, and δ wolves estimate the probable position of the prey (optimum solution). Grey wolves update their positions based on their distances from the prey. In order to emphasize exploration and exploitation during the search process, parameter a should decrease from 2 to 0. If $\left|\overrightarrow{A}\right| > 1$, the candidate solutions diverge from the prey; and if $|A| < 1$, the candidate solutions converge to the prey. This process continues and the GWO algorithm is terminated if the stopping criteria are satisfied. To understand how the GWO algorithm solves optimization problems theatrically, some notes can be summarized as follows:

- The concept of social hierarchy in the GWO algorithm helps grade the solutions and save the best solutions up to the current iteration.
- The encircling mechanism defines a 2D circle-shaped neighbor and the solution (in higher dimensions, the 2D circle can be extended to a 3D hyper-sphere).
- The random parameters (A and C) help grey wolves (candidate solutions) to define different hyper-spheres with random radii.
- The hunting approach implemented in the GWO algorithm allows grey wolves (candidate solutions) to locate the probable position of the prey (optimum solution).
- The adaptive values of parameters A and a guarantee exploration and exploitation in the GWO algorithm and also allow it to easily transfer between exploration and exploitation.
- By decreasing the values of A, a half of iterations are assigned to exploration $\left(\left|\overrightarrow{A}\right| > 1\right)$ and the other half of iterations are assigned to exploitation ($|A| < 1$).
- a and C are two main parameters of the GWO algorithm.

Figure 9.5 shows the flowchart of the GWO algorithm with details on the optimization process.

Fig. 9.5 Flowchart of the
GWO algorithm

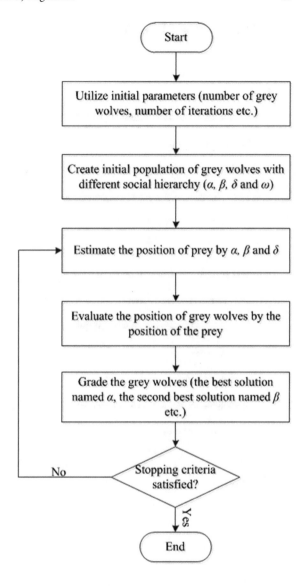

9.5   Pseudocode of GWO

Begin

Initialize the population of grey wolves X_i ($i = 1, 2, \dots, n$)

Initialize a, A, and C

Calculate the fitness values of search agents and grade them. (X_α= the best solution in the search agent, X_β= the second best solution in the search agent, and X_δ= the third best solution in the search agent)

$t = 0$

While ($t <$ Max number of iterations)

    For each search agent

        Update the position of the current search agent by Equation (9.7)

    End for

Update a, A, and C

Calculate the fitness values of all search agents and grade them

Update the positions of X_a, X_β, and X_δ

$t = t+1$

End while

End

9.6   Conclusions

This chapter described the grey wolf optimization (GWO) algorithm as one of the new meta-heuristic algorithms. The GWO algorithm was inspired by the lift style of the pack of grey wolves (social hierarchy and hunting mechanism). Also, this chapter presented a brief literature review of GWO, described the natural process of grey wolves' life style and the mathematical equations of GWO, and finally presented a pseudocode of GWO.

References

Gholizadeh, S. (2015). Optimal design of double layer grids considering nonlinear behaviour by sequential grey wolf algorithm. *Journal of Optimization in Civil Engineering, 5*(4), 511–523.

Mech, L. D. (1999). Alpha status, dominance, and division of labor in wolf packs. *Canadian Journal of Zoology, 77*(8), 1196–1203.

Mirjalili, S., Mirjalili, S. M., & Lewis, A. (2014). Grey wolf optimizer. *Advances in Engineering Software, 69*(2014), 46–61.

Mirjalili, S. (2015). How effective is the grey wolf optimizer in training multi-layer perceptron. *Applied Intelligence, 43*(1), 150–161.

Mirjalili, S. M., & Mirjalili, S. Z. (2015). Full optimizer for designing photonic crystal waveguides: IMoMIR framework. *IEEE Photonics Technology Letters, 27*(16), 1776–1779.

Mirjalili, S. M., Mirjalili, S., & Mirjalili, S. Z. (2015). How to design photonic crystal LEDs with artificial intelligence techniques. *Electronics Letters, 51*(18), 1437–1439.

Muro, C., Escobedo, R., Spector, L., & Coppinger, R. (2011). Wolf-pack (Canis Lupus) hunting strategies emerge from simple rules in computational simulations. *Behavioral Processes, 88*(3), 192–197.

Naderizadeh, M., & Baygi, S. J. M. (2015). Statcom with grey wolf optimizer algorithm based pi controller for a grid Connected wind energy system. *International Research Journal of Applied and Basic Sciences, 9*(8), 14–21.

Noshadi, A., Shi, J., Lee, W. S., Shi, P., & Kalam, A. (2015). Optimal PID-type fuzzy logic controller for a multi-input multi-output active magnetic bearing system. *Neural Computing and Applications, 27*(7), 1–16.

Saremi, S., Mirjalili, S. Z., & Mirjalili, S. M. (2015). Evolutionary population dynamics and grey wolf optimizer. *Neural Computing and Applications, 26*(5), 1257–1263.

Sulaiman, M. H., Mustaffa, Z., Mohamed, M. R., & Aliman, O. (2015). Using the grey wolf optimizer for solving optimal reactive power dispatch problem. *Applied Soft Computing, 32* (2015), 286–292.

Wong, L. I., Sulaiman, M. H., & Mohamed, M. R. (2015). Solving economic dispatch problems with practical constraints utilizing grey wolf optimizer. *Applied Mechanics and Materials, 785* (2015), 511–515. Trans Tech Publications.

Yusof, Y., & Mustaffa, Z. (2015). Time series forecasting of energy commodity using grey wolf optimizer. In *Proceedings of the international multiconference of engineers and computer scientists (IMECS 2015)*, Hong Kong, 18–20 March.

Chapter 10
Shark Smell Optimization (SSO) Algorithm

Sahar Mohammad-Azari, Omid Bozorg-Haddad and Xuefeng Chu

Abstract In this chapter, the shark smell optimization (SSO) algorithm is presented, which is inspired by the shark's ability to hunt based on its strong smell sense. In Sect. 10.1, an overview of the implementations of SSO is presented. The underlying idea of the algorithm is discussed in Sect. 10.2. The mathematical formulation and a pseudo-code are presented in Sects. 10.3 and 10.4, respectively. Section 10.5 is devoted to conclusion.

10.1 Introduction

Generally, all animals have abilities that ensure their survival in the nature. Some species have special abilities which distinguish them from others (Costa and Sinervo 2004). Finding the prey and the movement of hunter toward the prey are two important factors in the hunting process. Animals that are able to find the prey in a short time with a correct movement, are a successful hunter. Shark is one of the most well-known and superior hunter in the nature. The reason of this superiority is the shark's ability to find the prey in a short time based on its strong smell sense in a large search space.

Based on this shark's ability, Abedinia et al. (2014) developed a meta-heuristic algorithm named shark smell optimization (SSO), and evaluated its efficiency based

S. Mohammad-Azari · O. Bozorg-Haddad (✉)
Department of Irrigation and Reclamation Engineering, Faculty of Agricultural Engineering and Technology, College of Agriculture and Natural Resources, University of Tehran, 3158777871 Karaj, Iran
e-mail: OBHaddad@ut.ac.ir

S. Mohammad-Azari
e-mail: sahar.mazari@ut.ac.ir

X. Chu
Department of Civil and Environmental Engineering, North Dakota State University, Dept 2470, Fargo, ND 58108-6050, USA
e-mail: Xuefeng.Chu@ndsu.edu

© Springer Nature Singapore Pte Ltd. 2018
O. Bozorg-Haddad (ed.), *Advanced Optimization by Nature-Inspired Algorithms*,
Studies in Computational Intelligence 720, DOI 10.1007/978-981-10-5221-7_10

on some standard benchmark functions and compared with other meta-heuristic algorithms. Then, they investigated the performance of SSO in a real frequency control engineering problem in the electric power systems.

Abedinia and Amjadi (2015) applied a hybrid prediction model based on the neural network and chaotic SSO for wind power forecasting. The number of hidden nodes in a neural network was optimized by using the chaotic SSO model. In order to evaluate the efficiency of their proposed model, two case studies were considered and the results were compared with 14 other prediction methods. The results demonstrated the capability of their proposed model to cope with the variability and intermittency of wind power time series for providing accurate predictions.

Gnanaskaran et al. (2016) applied the SSO algorithm as an efficient method for finding the optimal size and location of shunt capacitors to minimize the cost due to energy loss and reactive power compensation of a distribution system. The results indicated the superiority of SSO compared with other classical algorithms. The reason of this superiority was that acquiring optimal solutions through simple formulation satisfied the problem constraints.

Ghaffari et al. (2016) investigated optimal economic load dispatch using the SSO algorithm. In their problem, risk constraints for unpredictable and uncertain behavior of wind were considered. In fact, they investigated the balance between cost and risk over a 30-bus power system with SSO and compared the results with those from other conventional methods.

Ehteram et al. (2016) evaluated the capability of SSO as a meta-heuristic optimization algorithm for operation of the single-reservoir (Bazoft) and multi-reservoir (Larson) systems. Then, the obtained results were compared with those from GA and the PSO algorithm based on certain performance criteria. The results indicated the superiority of SSO over two other algorithms due to the higher reliability and lower vulnerability. In addition, the application of the SSO algorithm was suggested for complex multi-objective reservoir systems with several operators.

10.2 Underlying Idea of SSO

The olfactory system in each animal is the primary sensory system which responds to the chemical signal from a remote source. In fishes, the smell receptors are located in the olfactory pits which are positioned on the sides of their heads. Each pit has two outside openings through which water flows in and out. The mechanism of water movement inside the pit is formed through the wave motion of tiny hair on the cells lining the pit and the force caused by the movement of fish in water. Dissolved chemicals connect to a pleated surface in the olfactory nerve endings (Abedinia et al. 2014). In vertebrates, unlike other sensory nerves, the olfactory receptors are directly connected to their brains without any nerve intermediaries.

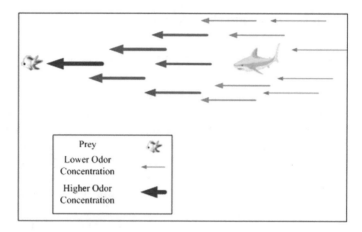

Fig. 10.1 Schematic of shark's movement toward the source of the smell

The smell impulses are received by the portion in the front of the brain called olfactory bulb. Fishes have two olfactory bulbs, each of which is located in an olfactory pit. The allocation of the larger surface of olfactory pits to smell nerves and larger smell information centers in the brain make the fishes' smell sense stronger (Magnuson 1979). Eels and sharks have the largest olfactory bulbs for smell information processing. About 400 million years ago, the first sharks appeared in the oceans as the superior hunters in the nature. One of the reasons of shark survival in the nature is its ability to capture the prey with its strong smell sense.

The shark's smell sense is one of its most effective senses. When a shark swims in water, the water flows through its nostrils which are located along the sides of its snout. After the water enters the olfactory pits, it flows through the folds of skin which is covered with sensory cells. Some sharks have this ability to detect the slightest trace of blood due to the sensory cells (Sfakiotakis et al. 1999). For example, a shark can detect one drop of blood in a large swimming pool. Accordingly, the shark can smell an injured fish from up to one kilometer away (Abedinia et al. 2014).

The shark's smell sense can consider as a guide for it. The smell that comes from the left side of the shark passes the left pit before entering the right pit. This process helps shark to find the source of the smell (Wu and Yao-Tsu 1971). The schematic of shark's movement toward the source of the smell is shown in Fig. 10.1.

In this movement, concentration plays an important role to guide the shark to its prey. In other words, a higher concentration results in a true movement of the shark. This characteristic is the base of development of an optimization algorithm to find the optimal solution of a problem.

10.3 Formulation of the SSO Algorithm

Some assumptions are considered to construct the mathematical formulation. They include:

(1) Fish is injured and this occurrence leads to blood injection into the sea (the search space). As a result, the velocity of the injured fish can be neglected compared against the shark movement velocity. In other words, the source (prey) is assumed to be fixed.

(2) Blood injection to sea water occurs regularly. The effect of water flow on distorting particles of odor is neglected. It is obvious that the odor particles are stronger near the injured fish. Consequently, following the odor particles helps shark to approach the prey.

(3) One injured fish results in one odor source in the search environment of the shark (Abedinia et al. 2014).

10.3.1 Initialization of SSO: Finding Initial Odor Particles

The search process begins when the shark smells odor. In fact, the particles of odor have a weak diffusion from an injured fish (prey). In order to model this process, a population of initial solutions are randomly generated for an optimization problem in the feasible search space. Each of these solutions represents a particle of odor which shows a possible position of the shark at the beginning of the search process.

$$\left[x_1^1, x_2^1, \ldots, x_{NP}^1 \right], \tag{10.1}$$

where x_i^1 = ith initial position of the population vector or ith initial solution; and NP = population size. The related optimization problem can be expressed as:

$$x_i^1 = \left[x_{i,1}^1, x_{i,2}^1, \ldots, x_{i,ND}^1 \right] \quad i = 1, 2, \ldots, NP, \tag{10.2}$$

where $x_{i,j}^1$ = jth dimension of the shark's ith position or jth decision variable of ith position of the shark $\left(x_i^1 \right)$; and ND = number of decision variables in the optimization problem.

The odor intensity at each position reflects its closeness to the prey. This process is modeled in the SSO algorithm through an objective function. Assuming a maximization problem and considering the general principle, a higher value of the objective function represents stronger odor (or more odor particles). Consequently, this process represents a closer position of the shark to its prey. The SSO algorithm initiates according to this view (Abedinia et al. 2014).

10.3.2   Shark Movement Toward the Prey

The shark at each position moves with a velocity to become closer to the prey. Based on the position vectors, the initial velocity vector can be expressed as:

$$\left[V_1^1, V_2^1, \ldots, V_{NP}^1\right] \tag{10.3}$$

In Eq. (10.3), the velocity vectors have components in each dimension.

$$V_i^1 = \left[V_{i,1}^1, V_{i,2}^1, \ldots, V_{i,ND}^1\right] \quad i = 1, \ldots, ND \tag{10.4}$$

The shark follows the odor and the direction of its movement is determined based on the intensity of odor. The velocity of the shark is increased due to the increased concentration of odor. From the optimization point of view, this movement is modeled mathematically by the gradient of the objective function. The gradient indicates the direction in which the function increases with the highest rate. Equation (10.5) shows this process (Abedinia et al. 2014):

$$V_i^k = \eta_k.R1.\nabla(\text{OF})|_{x_i^k} \quad i = 1, \ldots, \text{NP} \quad k = 1, \ldots, k_{\max}, \tag{10.5}$$

where V_i^k = velocity of the shark which is approximately constant; OF = objective function; ∇ = gradient of the objective function; $k_{\max}$ = maximum number of stages for forward movement of the shark; k = number of stages; η_k = a value in the interval [0,1]; and $R1$ = a random value which is uniformly distributed in the interval [0,1].

η_k is in the interval [0,1] as it may be impossible for a shark to reach the velocity determined by the gradient function. The parameter $R1$ gives more random search inherent to the SSO algorithm. The idea of considering $R1$ has been taken from the gravitational search algorithm (GSA). The velocity in each dimension can be calculated by Eq. (10.6) (Abedinia et al. 2014):

$$V_{i,j}^k = \eta_k.R1.\frac{\partial(\text{OF})}{\partial x_j}\bigg|_{x_{i,j}^k} \quad i = 1, \ldots, \text{NP } j = 1, \ldots, \text{ND } k = 1, \ldots, k_{\max} \tag{10.6}$$

Due to the existence of inertia, acceleration of the shark is limited and its velocity depends on its previous velocity. This process is modeled by a modified Eq. (10.6) as follows:

$$V_{i,j}^k = \eta_k . R1 . \left. \frac{\partial(\text{OF})}{\partial x_j} \right|_{x_{i,j}^k} + \alpha_k . R2 . V_{i,j}^{k-1}$$
$$i = 1, \ldots, \text{NP} \ j = 1, \ldots, \text{ND} \ k = 1, \ldots, k_{\max},$$
(10.7)

where α_k = rate of momentum or inertia coefficient that has a value in the interval of [0,1] and becomes a constant for stage k; and $R2$ = random number generator with a uniform distribution on the interval [0,1], which is intended for the momentum term. A larger value of α_k indicates higher inertia and hence the current velocity is more dependent on the previous velocity. From the mathematical point of view, the application of momentum leads to smoother search paths in the solution space. $R2$ increases the diversity of the search in the algorithm. For the velocity in the first stage $\left(V_{i,j}^1 \right)$, it is possible to neglect the initial velocity of the shark before starting the search process $\left(V_{i,j}^0 \right)$ or allocate a very small value to it.

The velocity of the shark can be increased up to a specified limit. Unlike most fishes, sharks do not have swim bladders to help them stay afloat. So, they cannot be static and must swim upward in a direction even with a low velocity. This process occurs using the strong tail fin which acts as a propulsion (Wu and Yao-Tsu 1971). The normal velocity of a shark is about 20 km/h which is increased up to 80 km/h when the shark tends to attack. The ratio of the highest to lowest velocities of the sharks is limited (For example, $\frac{80}{20} = 4$). The velocity limiter used for each stage of the SSO algorithm can be expressed as (Abedinia et al. 2014):

$$\left| V_{i,j}^k \right| = \min \left[\left| \eta_k . R1 . \left. \frac{\partial(\text{OF})}{\partial x_j} \right|_{x_{i,j}^k} + \alpha_k . R2 . V_{i,j}^{k-1} \right|, \left| \beta_k . V_{i,j}^{k-1} \right| \right],$$
$$i = 1, \ldots, \text{NP}, \quad j = 1, \ldots, \text{ND}, \quad k = 1, \ldots, k_{\max}$$
(10.8)

where β_k = velocity limiter ratio for stage k. The value of $V_{i,j}^k$ is calculated by Eq. (10.8) and it has the same sign as the term selected by the minimum operator in Eq. (10.8). Due to forward movement of the shark, its new position Y_i^{k+1} is determined based on its previous position and velocity:

$$Y_i^{k+1} = X_i^k + V_i^k . \Delta t_k \quad i = 1, \ldots, \text{NP} \quad k = 1, \ldots, k_{\max},$$
(10.9)

where Δt_k = time interval of stage k. Δt_k is assumed one for all stages for the purpose of simplicity. Each component of $V_{i,j}^k$ $(j = 1, \ldots, \text{ND})$ of vector V_i^k is obtained by Eq. (10.8).

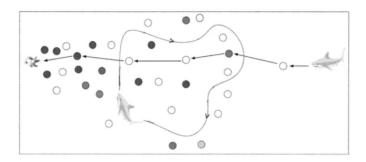

Fig. 10.2 Rotational movement of a shark

In addition to forward movement, the shark also has a rotational movement in its direction to find stronger odor particles. In fact, this improves its progress (Yao-Tsu 1971). The simulation of this movement is shown in Fig. 10.2.

As shown in Fig. 10.2, the rotational movement of the shark takes place along a closed contour which is not necessarily a circle. From the optimization view, in order to find better solutions the shark does a local search in each stage. This local search in the SSO algorithm is modeled by Abedinia et al. (2014):

$$Z_i^{k+1,m} = Y_i^{k+1} + R3.Y_i^{k+1} \quad m = 1,\ldots,M \quad i = 1,\ldots,NP \quad k = 1,\ldots,k_{\max},$$

$$(10.10)$$

where $Z_i^{k+1,m}$ = position of point m in the local search; $R3$ = a random number with a uniform distribution in the interval $[-1, 1]$; and M = number of points in the local search of each stage.

Since this operator does a local search around Y_i^{k+1}, the limit of $R3$ can be considered in the interval $[-1, 1]$. M points of the local search $Z_i^{k+1,m}$ are in the vicinity of Y_i^{k+1} (If the random number generator generates zero, Y_i^{k+1} is obtained). A closed contour is obtained by the connections of M points, which is similar to the rotational movement of the shark. During the rotational movement, if the shark finds a point with a stronger odor it follows the search from that point. This process is shown in Fig. 10.2. This characteristic in the SSO algorithm can be expressed as (Abedinia et al. 2014):

$$x_i^{k+1} = \arg\max\left\{ \text{OF}\left(Y_i^{k+1}\right), \text{OF}\left(Z_i^{k+1,i}\right),\ldots,\text{OF}\left(Z_i^{k+1,M}\right) \right\} \quad i = 1, 2,\ldots,NP$$

$$(10.11)$$

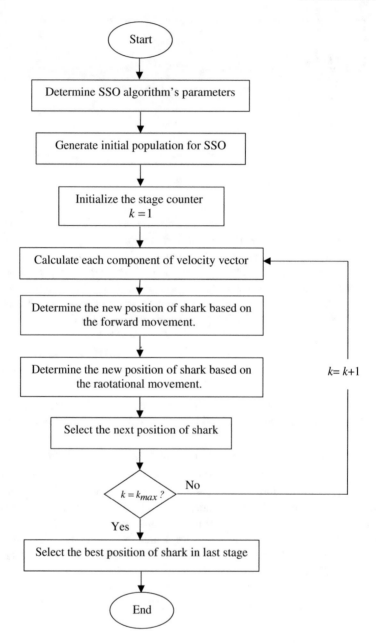

Fig. 10.3 Flowchart of the SSO algorithm

Table 10.1 Characteristics of the SSO algorithm

General algorithm	Gradient evolution algorithm
Decision variable	Odor particle which shows possible position of shark
Solution	Odor intensity
Old solution	Old position of shark
New solution	New position of shark
Best solution	Position of shark with the best fitness function
Fitness function	Odor particle intensity which represents the closer position of shark to prey
Initial solution	Randomly generated position of shark
Selection	Select shark's position based on the forward and rotational movements
Process of generating new solution	Forward movement and rotational movement

In Eq. (10.11), the objective function (*OF*) must be maximized. In other words, among Y_i^{k+1} obtained from forward movement and $Z_i^{k+1,m}$ (*m* = 1, 2, ..., *M*) obtained from the rotational movement, a solution with the highest objective function is selected as the next position of the shark (x_i^{k+1}). The cycle of forward and rotational movement will continue until *k* is equal to k_{max}.

Like other meta-heuristic optimization methods, the SSO algorithm also has a number of parameters that must be defined by users, including NP, k_{max}, η, α, and β in each stage. Changing these parameters during the SSO evolution based on an adaptive mechanism is an effective method in the applications. For example, such a mechanism may start adaptively from larger values of η and β and a smaller value of α, and then the values of η and β will be decreased and the value of α will be increased. So, in the initial stage of the evolution process, the algorithm will continue with large steps in order to enhance the search ability and for the last stage (when the algorithm approaches the optimal solution) the steps will be smaller to increase the resolution of search around the optimal solution. After setting the parameters, the population and the stage counter of the SSO are initialized.

Population will be evolved by the operators of forward and rotational movements. Finally, the best solution in the final stage is selected for the optimization problem. The search operators in the SSO algorithm including the gradient-based forward movement and local search based rotational movement are not used in any other meta-heuristic algorithms (Abedinia et al. 2014).

The flowchart of the SSO algorithm is shown in Fig. 10.3, and the parameters and variables used in the SSO algorithm are listed in Table 10.1.

10.4   Pseudo-Code of SSO

Begin

    Step 1. Initialization

        Set parameters NP, k_{max}, η_k, α_k, and β_k $\;\left(k=1,\,2,\,...,\,k_{max}\right)$

        Generate initial population with all individuals

        Generate each decision randomly within the allowable range

        Initialize the stage counter $k = 1$

  For $k = 1 : k_{max}$

    Step 2. Forward movement

        Calculate each component of the velocity vector, $v_{i,j}\left(i = 1, ..., NP\,,\, j = 1,..., ND\right)$

        Obtain new position of shark based on forward movement, $Y_i^{k+1}\left(i = 1, ..., NP\right)$

    Step 3. Rotational movement

        Obtain position of shark based on rotational movement, $Z_i^{k+1,m}\left(m = 1, .., M\right)$

        Select next position of shark based on the two movements $X_i^{k+1}\left(i = 1, ..., NP\right)$

  End for k

    Set $k = k+1$

    Select the best position of shark in the last stage which has the highest OF value

End

10.5   Conclusion

In this chapter, the shark smell optimization (SSO) algorithm was introduced as one of the new meta-heuristic optimization methods. This algorithm was developed based on the hunting ability of sharks to use their smell sense. This is a stochastic search optimization algorithm which initiates with a sets of random solutions and continues the search to find the optimal solution. In fact, this algorithm applies a gradient-based forward movement and a local search based rotational movement during the optimization process. In this chapter, the SSO algorithm was introduced and its mathematical formulation was presented.

References

Abedinia, O., & Amjadi, N. (2015). Short-Term wind power prediction based on hybrid neural network and chaotic shark smell optimization. *International Journal of Precision Engineering and Manufacturing-Green Technology, 2*(3), 245–254.

Abedinia, O., Amjady, N., & Ghasemi, A. (2014). A new metaheuristic algorithm based on shark smell optimization. *Complexity.* doi:10.1002/cplx.21634

Costa, D. P., & Sinervo, B. (2004). Field physiology: physiological insights from animals in nature. *Annual Review of Physiology, 66,* 209–238.

Ehteram, M., Karimi, H., Musavi, S. F., & EL-Shafie, A. (2017). *Optimizing dam and reservoirs operation based model utilizing shark algorithm approach.* Knowledge-Based Systems (In Press). doi:10.1016/j.knosys.2017.01.026

Ghaffari, S., Aghajani, Gh, Noruzi, A., & Hedayati-Mehr, H. (2016). Optimal economic load dispatch based on wind energy and risk constrains through an intelligent algorithm. *Complexity, 21*(S2), 494–506.

Gnanasekaran, N., Chandramohan, S., Sathish Kumar, P., & Mohamed Imran, A. (2016). Optimal placement of capacitors in radial distribution system using shark smell optimization algorithm. *Ain Shams Engineering Journal, 7,* 907–916.

Magnuson, J. J. (1979). 4 Locomotion by Scombrid fishes: hydromechanics, morphology and behavior. *Fish Physiology, 7,* 239–313.

Sfakiotakis, M., Lane, D. M., & Davies, J. B. C. (1999). Review of fish swimming modes for aquatic locomotion. *IEEE Journal of Oceanic Engineering, 24,* 237–252.

Wu, T. Yao-Tsu. (1971). Hydromechanics of swimming propulsion. Part 1. Swimming of a two-dimensional flexible plate at variable forward speeds in an inviscid fluid. *Journal of Fluid Mechanics, 46*(2), 337–355.

Chapter 11
Ant Lion Optimizer (ALO) Algorithm

Melika Mani, Omid Bozorg-Haddad and Xuefeng Chu

Abstract This chapter introduces the ant lion optimizer (ALO), which mimics the hunting behavior of antlions in the larvae stage. Specifically, this chapter includes literature review, details of the ALO algorithm, and a pseudo-code for its implementation.

11.1 Introduction

Mirjalili (2015) introduced the ant lion optimizer (ALO) algorithm and proved its capability by solving 19 different mathematical benchmark problems and three classical engineering problems including three-bar truss design, cantilever beam design, and gear train design. In addition, ALO was used for optimizing the shape of two ship propeller as a challenging constrained problem with a diverse search space, which showed the ability of the ALO algorithm for solving real complex problems. Yamany et al. (2015) used ALO for determining weights and biases in a training process of multilayer perceptron (MLP) for having a minimum error and an appropriate classification rate. In the research, the performance of ALO was compared with those of genetic algorithm (GA), particle swarm optimization (PSO) algorithm, and ant colony optimization (ACO) algorithm to show its capability. Zawbaa et al. (2015) applied ALO to an optimization problem of feature

M. Mani · O. Bozorg-Haddad (✉)
Department of Irrigation and Reclamation Engineering, Faculty of Agricultural Engineering and Technology, College of Agriculture and Natural Resources, University of Tehran, 31587-77871 Karaj, Tehran, Iran
e-mail: OBHaddad@ut.ac.ir

M. Mani
e-mail: Melika.Mani@ut.ac.ir

X. Chu
Department of Civil and Environmental Engineering, North Dakota State University, Dept 2470, 58108-6050 Fargo, ND, USA
e-mail: Xuefeng.Chu@ndsu.edu

© Springer Nature Singapore Pte Ltd. 2018
O. Bozorg-Haddad (ed.), *Advanced Optimization by Nature-Inspired Algorithms*,
Studies in Computational Intelligence 720, DOI 10.1007/978-981-10-5221-7_11

selection to maximize the classification performance. They compared ALO with PSO and GA, and demonstrated the better performance of ALO. Ali et al. (2016) employed ALO to optimize allocation and sizing of distributed generation (DG), which contained photovoltaic and wind turbines. The results showed the superiority of ALO over several other optimization algorithms such as GA and PSO. Dubey et al. (2016) applied ALO for solving a hydrothermal power generation scheduling (HTPGS) problem with wind integration, which was a nonlinear, non-convex, and highly complex optimization problem. The results showed high ability of ALO for finding powerful solutions in a complex decision space.

Talatahari (2016) implemented ALO for an optimal design of skeletal structures. The efficiency of ALO was illustrated by comparing with various classical optimization algorithms. Zawbaa et al. (2016) enhanced the performance of ALO for a feature selection problem by developing a "chaotic" version of ALO. Kamboj et al. (2016) used ALO for a non-convex optimization problem of economic load dispatch of electric power systems. Petrović et al. (2016) employed ALO for solving a combinatorial optimization problem of integrated planning and scheduling. Raju et al. (2016) presented an application of ALO for simultaneous optimization of different controllers. Saxena and Kothari (2016) used ALO for solving antenna array synthesis and other electromagnetic optimization problems. Yao and Wang (2016) developed the dynamic adaptive ant lion optimizer (DAALO) by replacing random walk of ants with levy flight and adjusting size of traps. The results from the unmanned aerial vehicle (UAV) route planning problem showed high efficiency, robustness, and feasibility of DAALO. Mirjalili et al. (2017) developed multi-objective ALO (MOALO) and used standard unconstrained and constraint test functions to show the efficiency of the algorithm. Also, MOALO was applied to solve several engineering design problems including cantilever beam design, brushless dc wheel motor design, disk brake design, 4-bar truss design, safety isolating transformer design, speed reduced design, and welded beam design. The results showed the capability of the algorithm in solving challenging real-world engineering problems. Kaur and Mahajan (2016) applied ALO for optimization of the communities in large networks. Rajan et al. (2017) applied ALO to determine optimal reactive power dispatch of systems, which was a highly nonlinear, non-convex challenging optimization problem. They modified ALO by introducing a noble weighted elitism concept in the elitism phase of the original ALO. Kaushal and Singh (2017) used ALO to optimize the allocation of stocks in portfolio and compared with GA, showing that ALO outperformed GA for portfolio designing.

11.2   Mapping Antlions Hunting Mechanism into the ALO

Antlions are in the family of Myrmeleontidae and belong to the order of Neuroptera. The life cycle of antlions includes two main phases, larva and adult, which takes 2 to 3 years. The antlions' life cycle mostly occurs in larvae and adulthood has only 3 to 5 weeks. The larvae of antlions are also known as

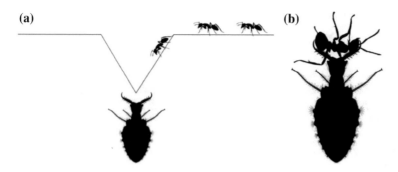

Fig. 11.1 Antlion hunting behavior

doodlebugs, which have a predatory habit. Adult antlions, which are less well known, can fly and maybe are mistakenly identified as dragonflies or damselflies. The name of "antlions" best describes their unique hunting behavior and their favorite prey which is ants. The larvae of some antlions species dig cone-shaped pits with different sizes and wait at the bottom of the pits for ants or other insects to slip on the loose sands and fall in, as shown in Fig. 11.1a.

When an insect is in a trap, the antlion will try to catch it while the trapped insect will try to escape. The antlion intelligently tries to slide the prey into the bottom of the pit by throwing sands toward the edge of the pit. After catching the prey, the antlion pulls it under the soil and consumes it (Fig. 11.1b). After feeding is completed, the antlion flicks the leftovers of the prey out of the pit and prepares the pit for next hunting.

It should be noted that the size of the antlion's trap depends on the level of antlion hunger and the shape of the moon. Antlions dig larger pits when they become hungry and also when the moon is full (Goodenough 2009). For larger pits, the chance of successful hunting increases. The ALO algorithm is inspired by this intelligent hunting behavior of antlions and the interaction with their favorite prey, ants. So the main steps of antlions' hunting are mathematically modeled in the ALO algorithm.

In the ALO algorithm, ants are search agents and move over the decision space, and antlions are allowed to hunt them and become fitter. In each iteration, the position of each ant is updated with respect to the selected antlion based on roulette wheel and elite (best antlion obtained so far). By the roulette wheel selection operator, solutions with the better fitness function have more chance to be selected as the antlion with a larger trap has more chance to hunt more ants. Table 11.1 lists the characteristics of the ALO algorithm.

In the ALO algorithm, the first positions of antlions and ants are initialized randomly and their fitness functions are calculated. Then, the elite antlion is determined. In each iteration for each ant, one antlion is selected by the roulette wheel operator and its position is updated with the aid of two random walk around the roulette selected antlion and elite. The new positions of ants are evaluated by calculating their fitness functions and comparing with those of antlions. If an ant becomes fitter than its corresponding antlion, its position is considered as a new position for the antlion in the

Table 11.1 Characteristics of the ALO algorithm

General algorithm	Antlion optimizer
Decision variable	Antlion's and Ant's positions in each dimension
Solution	Antlion's position
Old solution	Old position of antlion
New solution	New position of antlion
Best solution	Elite antlion
Fitness function	Desirability of elite
Initial solution	Random antlion
Selection	Roulette wheel
Process of generating new solution	Random walk over the decision space

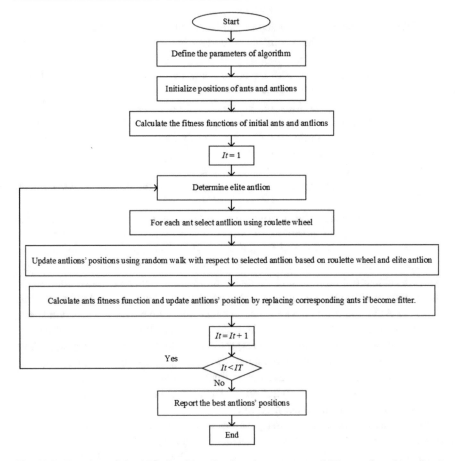

Fig. 11.2 Flowchart of the ALO algorithm (It = iteration counter; and IT = number of iterations)

next iteration. Also, the elite will be updated if the best antlion achieved in the current iteration becomes fitter than the elite. These steps are repeated until the end of iterations. The flowchart of the ALO algorithm is illustrated in Fig. 11.2.

11.2.1 Initialization of Positions of Ants and Antlions and Evaluation of Their Fitness Functions

In the ALO algorithm, there are two populations, ants and antlions. As afore-mentioned, ants are the search agents in the decision space and antlions which hide somewhere in the decision space, can hunt them, and catch their positions to become fitter. In an optimization problem with N decision variables (N-dimensional optimization problem), the ants/antlions' positions in an N-dimensional space are the decision variables. So each dimension of the ant/antlion's position belongs to one of the decision variables which can be expressed as:

$$\text{Ant's location} = [A_1, A_2, \ldots, A_n, \ldots, A_N] \tag{11.1}$$

$$\text{Antlion's location} = [Al_1, Al_2, \ldots, AL_n, \ldots, Al_N], \tag{11.2}$$

where $A_n/AL_n = n$th decision variable; and N = number of decision variables.

The ALO algorithm starts with randomly generated matrices of the positions of ants and antlions within their specified ranges as follow:

$$P_{\text{ant}} = \left\{ \begin{matrix} A_{1,1} & A_{1,2} & \cdots & A_{1,n} & \cdots & A_{1,N} \\ A_{2,1} & A_{2,2} & \cdots & A_{2,n} & \cdots & A_{1,N} \\ \vdots & \vdots & \vdots & \vdots & \vdots & \vdots \\ A_{m,1} & A_{m,2} & \vdots & A_{m,n} & \vdots & A_{m,N} \\ \vdots & \vdots & \vdots & \vdots & \vdots & \vdots \\ A_{M,1} & A_{M,2} & \cdots & A_{M,n} & \cdots & A_{M,N} \end{matrix} \right\}, \tag{11.3}$$

where P_{ant} = matrix of ants' positions; $A_{m,n} = n$th decision variable of the mth ant; and M = number of ants.

$$P_{\text{antlion}} = \left\{ \begin{matrix} Al_{1,1} & Al_{1,2} & \cdots & Al_{1,n} & \cdots & Al_{1,N} \\ Al_{2,1} & Al_{2,2} & \cdots & Al_{2,n} & \cdots & Al_{1,N} \\ \vdots & \vdots & \vdots & \vdots & \vdots & \vdots \\ Al_{m,1} & Al_{m,2} & \vdots & Al_{m,n} & \vdots & Al_{m,N} \\ \vdots & \vdots & \vdots & \vdots & \vdots & \vdots \\ Al_{M,1} & Al_{M,2} & \cdots & Al_{M,n} & \cdots & Al_{M,N} \end{matrix} \right\}, \tag{11.4}$$

where P_{antlion} = matrix of antlions' positions; and $Al_{m,n} = n$th decision variable of the mth antlion.

For evaluating the ants and antlions, a fitness function is utilized and their fitness values are calculated during optimization and saved in the following matrices. In this process, the best antlion (antlion with the best fitness) is selected as elite.

$$F_{\text{ant}} = \begin{cases} f([A_{1,1} & A_{1,2} & \cdots & A_{1,n} & \cdots & A_{1,N}]) \\ f([A_{2,1} & A_{2,2} & \cdots & A_{2,n} & \cdots & A_{1,N}]) \\ \vdots & \vdots & \vdots & \vdots & \vdots & \vdots \\ f([A_{m,1} & A_{m,2} & \vdots & A_{m,n} & \vdots & A_{m,N}]) \\ \vdots & \vdots & \vdots & \vdots & \vdots & \vdots \\ f([A_{M,1} & A_{M,2} & \cdots & A_{M,n} & \cdots & A_{M,N})] \end{cases}, \qquad (11.5)$$

where F_{ant} = matrix of the ants' fitness functions; and f = fitness function.

$$F_{\text{antlion}} = \begin{cases} f(Al_{1,1} & Al_{1,2} & \cdots & Al_{1,n} & \cdots & Al_{1,N}) \\ f(Al_{2,1} & Al_{2,2} & \cdots & Al_{2,n} & \cdots & Al_{1,N}) \\ \vdots & \vdots & \vdots & \vdots & \vdots & \vdots \\ f(Al_{m,1} & Al_{m,2} & \vdots & Al_{m,n} & \vdots & Al_{m,N}) \\ \vdots & \vdots & \vdots & \vdots & \vdots & \vdots \\ f(Al_{M,1} & Al_{M,2} & \cdots & Al_{M,n} & \cdots & Al_{M,N}) \end{cases}, \qquad (11.6)$$

where F_{antlion} = matrix of the antlions' fitness functions.

11.2.2 Digging Trap

In this step, by using the roulette wheel operator for each ant, an antlion is selected. It should be noted that in the ALO algorithm each ant can fall into only one antlion trap in each iteration. The roulette wheel selected antlion for each ant is the one that has trapped the ant. By using the roulette wheel operator, the solution with a better fitness function has more chance to be selected, as an antlion with a larger trap can hunt more ants.

11.2.3 Sliding Ants Toward Antlion

When an ant falls into the trap, the antlion starts shooting sand outward the center of the pit for sliding down the ant which is trying to escape. This behavior is mathematically modeled by shrinking the radius of the ant's random walk. So the range of boundary for all decision variables is decreased and updated, as expressed in Eqs. (11.7) and (11.8) (Mirjalili 2015).

$$\gamma(It) = \frac{c(It)}{R} \tag{11.7}$$

$$\delta(It) = \frac{d(It)}{R}, \tag{11.8}$$

where $\gamma(It)$ = modified vector including the minimum of all decision variables at the Itth iteration; $c(It)$ = vector including the minimum of all decision variables at the Itth iteration; R = a ratio given by Eq. (11.9); $\delta(It)$ = modified vector including the maximum of all decision variables at the Itth iteration; and $d(It)$ = vector including the maximum of all decision variables at the Itth iteration (Mirjalili 2015).

$$R = 10^w \frac{It}{IT}, \tag{11.9}$$

where w = a constant defined based on the iteration number given by:

$$w = \begin{cases} 2 & \text{If } It > 0.1IT \\ 3 & \text{If } It > 0.5IT \\ 4 & \text{If } It > 0.75IT \\ 5 & \text{If } It > 0.9IT \\ 6 & \text{If } It > 0.95IT \end{cases} \tag{11.10}$$

When the iteration number in Eq. (11.10) increases, the radius of random walk decreases, which guarantees convergence of the ALO algorithm.

11.2.4 Entrapping Ants Inside Pits

Antlion traps affect the random walk of ants. In order to mathematically model this behavior, the boundary of ant random walk is adjusted in each iteration so that the ant moves in a hyper-sphere around the selected antlion trap. The lower and upper bounds of the ant random walk for each dimension in each iteration can be calculated by the following equations (Mirjalili 2015):

$$\gamma_m(It) = \text{Antlion}_l(It) + \gamma(It) \tag{11.11}$$

$$\delta_m(It) = \text{Antlion}_l(It) + \delta(It), \tag{11.12}$$

where $\gamma_m(It)$ = vector including the minimum of all decision variables for the mth ant in the Itth iteration; $\text{Antlion}_l(It)$ = position of the selected lth antlion in the Itth iteration; and $\delta_m(It)$ = vector including the minimum of all decision variables for the mth ant. Equations (11.11) and (11.12) show that ants' random walk is in the hyper-sphere, which is defined by vectors γ and δ around the roulette wheel selected antlion.

11.2.5 Random Walk of Ants

As ants move randomly in the nature for food, random walk is used for modeling their movement, which can be expressed as the following equation (Mirjalili 2015):

$$X = [0, \text{cumsum}(2r(1) - 1), \text{cumsum}(2r(2) - 1), \ldots, \text{cumsum}(2r(It) \\ - 1), \ldots, \text{cumsum}(2r(IT) - 1)], \tag{11.13}$$

where X = vector of random walk positions; $cumsum$ = cumulative sum; and $r(It)$ = stochastic function which is calculated by:

$$r(It) = \begin{cases} 1 & \text{if rand} > 0.5 \\ 0 & \text{otherwise} \end{cases}, \tag{11.14}$$

where $rand$ = random value generated with a uniform distribution between 0 and 1.

Figure 11.3 shows three random walks in 100 iterations. From Fig. 11.3, completely different behaviors of random walk can be observed.

To keep the random walks within the decision space, they are normalized by the min-max normalization method (Mirjalili 2015):

$$Z_n(It) = \frac{(X_n(It) - \alpha_n) + (\delta_n(It) - \gamma_n(It))}{(\beta_n - \alpha_n)}, \tag{11.15}$$

where $Z_n(It)$ = normalized random walk position of the nth decision variable at the Itth iteration; $X_n(It)$ = random walk position of the nth decision variable at the Itth iteration before normalization; α_n = minimum of random walks for the nth decision variable; β_n = maximum of random walks for the nth decision variable; $\gamma_n(It)$ = minimum of the nth decision variable at the Itth iteration; and $\delta_n(It)$ = maximum of the nth decision variable at the Itth iteration.

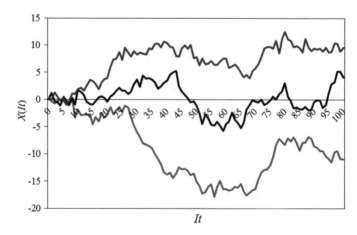

Fig. 11.3 Three random walk curves in one dimension started at zero

11.2.6 Elitism

Elitism is an important aspect of the evolutionary algorithms which allows them to keep the best solution during the optimization process. In the ALO algorithm, the best antlion in each iteration is saved as the elite antlion and the average of both random walks around the roulette wheel selected antlion and the elite is considered for generating new positions of ants, as expressed in Eq. (11.16).

$$\text{Ant}_m^{It} = \frac{R_l^{It} + R_e^{It}}{2},$$ (11.16)

where Ant_m^{It} = position of the selected mth antlion in the Itth iteration; R_l^{It} = random walk around the lth roulette wheel selected antlion at the Itth iteration; and R_e^{It} = random walk around the elite antlion at the Itth iteration.

11.2.7 Catching Prey and Reconstruct the Trap

At the final step of antlion hunting, an ant falls into the bottom of the trap and is caught by the antlion's jaw. The antlion pulls the ant into sand and consumes it. In the ALO algorithm, catching the prey occurs when an ant's fitness function becomes better than its corresponding antlion. In this situation, the antlion changes its position to the position of the hunted ant. This process can be mathematically expressed as:

$$\text{Antlion}_l^{It} = \text{Ant}_m^{It} \quad \text{If } f(\text{Ant}_m^{It}) > f(\text{Antlion}_l^{It}).$$ (11.17)

11.3 Termination Criteria

In the evolutionary algorithms, at the end of each iteration the termination criterion is applied to decide if the algorithm is to be stopped or continues the next iteration. A good termination criterion should guarantee convergence of the algorithm. To meet this purpose in the ALO algorithm by increasing the iteration number, the radius of random walk decreases and the maximum number of iterations is considered as a termination criterion.

11.4 User-Defined Parameters of the ALO Algorithm

In the ALO algorithm, only the number of search agents and the number of iterations are user-defined parameters. So, one of the main advantages of the ALO algorithm is that it has very few parameters to be adjusted. Generally, the evolutionary algorithms require a large number of user-defined parameters and the optimal settings of these parameters (to find an appropriate solution in a reasonable time) are different for each problem. Determining a good parameter setting often requires numerous careful, time-consuming experiments. For the ALO algorithm, however, only two user-defined parameters are required.

11.5   Pseudo-Code of the ALO Algorithm

Begin

    Input parameters of the algorithm and initial data

    Generate the initial population of ants and antlions randomly

    Calculate the fitness function values for ants and antlions

    Select the fittest antlion as the elite

    While the iteration number is smaller than the maximum iteration number

        For each ant

            Select an antlion by utilizing the roulette wheel operator

            Update the lower and upper bounds by using Equations (11.7) and (11.8)

            Update γ and δ using Equations (11.11) and (11.12)

            Generate two random walks around the roulette selected antlion and the elite and normalize them by Equation (11.15)

            Update the ant positions by Equation (11.16)

        End for

        Calculate the fitness function values for all ants and replace them in the corresponding matrix

        Replace an antlion with its corresponding ant if it becomes fitter [Equation (11.17)]

        Update the elite value if an antlion's fitness function value becomes better than the elite

    End while

    Report the best solution (elite)

End

11.6 Conclusion

This chapter introduced the antlion optimizer (ALO), which is inspired by the hunting behavior of antlions. In this chapter, after a brief literature review of the ALO algorithm and an explanation of the hunting behavior of antlions, the algorithmic fundamentals of the ALO algorithm are detailed and a pseudo-code is also presented.

References

Ali, E. S., Elazim, S. A., & Abdelaziz, A. Y. (2016). Optimal allocation and sizing of renewable distributed generation using Antlion optimization algorithm. *Electrical Engineering*. doi:10.1007/s00202-016-0477-z

Dubey, H. M., Pandit, M., & Panigrahi, B. K. (2016). Antlion optimization for short-term wind integrated hydrothermal power generation scheduling. *International Journal of Electrical Power & Energy Systems, 83*(1), 158–174.

Goodenough, J., McGuire, B., & Jakob, E. (2009). *Perspectives on animal behavior* (3rd ed.). New York, USA: John Wiley and Sons.

Kamboj, V. K., Bhadoria, A., & Bath, S. K. (2016). Solution of non-convex economic load dispatch problem for small-scale power systems using Antlion optimizer. *Neural Computing and Applications, 25*(5), 1–12.

Kaur, M. & Mahajan, A. (2016). Community Detection in Complex Networks: A Novel Approach Based on Antlion Optimizer. *Sixth International Conference on Soft Computing for Problem Solving, Punjab, India*. December 23–24.

Kaushal, K., & Singh, S. (2017). Allocation of stocks in a portfolio using Antlion algorithm: Investor's perspective. *IUP Journal of Applied Economics, 6*(1), 34–49.

Mirjalili, S. (2015). The Ant lion optimizer. *Advances in Engineering Software, 83*(1), 80–98.

Mirjalili, S., Jangir, P., & Saremi, S. (2017). Multi-objective Antlion optimizer: a multi-objective optimization algorithm for solving engineering problems. *Applied Intelligence, 46*(1), 79–95.

Petrović, M., Petronijević, J., Mitić, M., Vuković, N., Miljković, Z., & Babić, B. (2016). The Antlion optimization algorithm for integrated process planning and scheduling. *Applied Mechanics and Materials, 834*(1), 187–192.

Rajan, A., Jeevan, K., & Malakar, T. (2017). Weighted elitism based Antlion optimizer to solve optimum VAr planning problem. *Applied Soft Computing, 55*(1), 352–370.

Raju, M., Saikia, L. C., & Sinha, N. (2016). Automatic generation control of a multi-area system using Antlion optimizer algorithm based PID plus second order derivative controller. *International Journal of Electrical Power & Energy Systems, 80*(1), 52–63.

Saxena, P., & Kothari, A. (2016). Antlion optimization algorithm to control side lobe level and null depths in linear antenna arrays. *AEU-International Journal of Electronics and Communications, 70*(9), 1339–1349.

Talatahari, S. (2016). Optimum design of skeletal structures using Antlion optimizer. *International Journal of Optimization in Civil Engineering, 6*(1), 13–25.

Yamany, W. et al. (2015, September 20–22). A new multi-layer perceptrons trainer based on Antlion optimization algorithm. *Fourth International Conference on Information Science and Industrial Applications, Beijing, China.*

Yao, P., & Wang, H. (2016). Dynamic adaptive Antlion optimizer applied to route planning for unmanned aerial vehicle. *Soft Computing*. doi:10.1007/s00500-016-2138-6

Zawbaa, H. M., Emary, E., & Grosan, C. (2016). Feature selection via chaotic antlion optimization. *PLoS ONE, 11*(3), e0150652.

Zawbaa, H. M., Emary, E., & Parv, B. (2015, November 23–25). "Feature selection based on antlion optimization algorithm." *Third World Conference on Complex Systems, Marrakech, Morocco*.

Chapter 12
Gradient Evolution (GE) Algorithm

Mehri Abdi-Dehkordi, Omid Bozorg-Haddad and Xuefeng Chu

Abstract In this chapter, a meta-heuristic optimization algorithm named gradient evolution (GE) is discussed, which is based on a gradient search method. First, the GE algorithm and the underlying idea are introduced and its applications in some studies are reviewed. Then, the mathematical formulation and a pseudo-code of GE are discussed. Finally, the conclusion is presented.

12.1 Introduction

The gradient evolution (GE) algorithm is a meta-heuristic optimization algorithm which is derived from a gradient-based optimization method. A gradient defines the curve of a function at each point. Negative, zero, and positive gradients indicate the decreasing, flat, and increasing functions, respectively. Accordingly, the optimal solution is located at certain extreme point. This concept is the basis of the gradient-based search methods including Newton, quasi-Newton, and conjugate direction methods. Many researchers have applied the gradient-based methods or their combinations with other methods in solving optimization problems. The non-differentiable functions in complex problems often limit the application of the gradient-based methods. Kuo and Zulvia (2015) developed the GE algorithm, a novel meta-heuristic optimization algorithm which applied a modified gradient-based

M. Abdi-Dehkordi · O. Bozorg-Haddad (✉)
Department of Irrigation and Reclamation Engineering, Faculty of Agricultural Engineering
and Technology, College of Agriculture and Natural Resources, University of Tehran,
3158777871 Karaj, Iran
e-mail: OBHaddad@ut.ac.ir

M. Abdi-Dehkordi
e-mail: abdi.dehkordi@ut.ac.ir

X. Chu
Department of Civil and Environmental Engineering, North Dakota State University,
Dept 2470, Fargo, ND 58108-6050, USA
e-mail: Xuefeng.Chu@ndsu.edu

© Springer Nature Singapore Pte Ltd. 2018   117
O. Bozorg-Haddad (ed.), *Advanced Optimization by Nature-Inspired Algorithms*,
Studies in Computational Intelligence 720, DOI 10.1007/978-981-10-5221-7_12

method as the main updating rule. The GE algorithm explored the search space of an optimization problem using a set of vectors. Kuo and Zulvia also considered a set of operators in order to enhance the ability of their model in finding the optimal solution. They further evaluated the performance of the GE algorithm by using 15 benchmark test functions in three stages. In the first stage, the effects of changing parameters on the obtained results of the GE algorithm were investigated and the best parameter setting was determined. Then, the results of the GE algorithm were compared with those from other meta-heuristic algorithms including particle swarm optimization (PSO), differential evolution (DE), continuous genetic algorithm (GA), and artificial bee colony (ABC). The results indicated the better performance of the GE algorithm than those of the other meta-heuristic algorithms.

Kuo and Zulvia (2016) also proposed a K-means clustering algorithm based on the GE algorithm in order to derive the hidden stored information in data sets. The reason of development of this new algorithm was the dependency of the K-means algorithm on the initial centroids of clusters. In the proposed algorithm, the GE algorithm was utilized to find a good center for the K-means algorithm. The algorithm was validated by a number of benchmark datasets and the obtained results were compared with those from other meta-heuristic based K-means algorithms, indicating the superiority of the GE based K-means algorithm over the other meta-heuristic algorithms. Kuo and Zulvia (2016) finally proposed to consider the similarities between clusters as well as the similarities within clusters in a multi-objective structure.

12.2 Underlying Idea of the GE Algorithm

The GE algorithm is a population-based optimization method and its main rule is updated by the gradient search methods (including the Newton–Raphson method and Taylor series expansion), so it can be considered as a center differencing approach. Consequently, this algorithm can be applied to different optimization problems with differentiable and non-differentiable functions (Kuo and Zulvia 2015).

12.2.1 Gradient

If $f(x)$ is a differentiable function at point x, changing x causes a change in f (x) (Larson et al. 2007). In a single-variable function, the derivative is defined as the slope of the tangent or zero gradient at point x. The gradient of function $f(x)$ at a point like x_0 is determined by a secant line between $f(x_0)$ and $f(x_0 + \Delta x)$ (Fig. 12.1).

Fig. 12.1 Gradient
determination

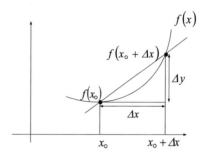

Thus, the gradient is determined by:

$$m = \frac{\Delta y}{\Delta x} = \frac{f(x_0 + \Delta x) - f(x_0)}{\Delta x} \tag{12.1}$$

where m = gradient; and Δy = change of $f(x)$ between points x_0 and $x_0 + \Delta x$.

The values of $f(x_0 + \Delta x)$ and $f(x_0)$ become closer for a smaller value of Δx, which increases the accuracy of calculation. So if Δx approaches zero, the secant line is tangent of $f(x)$ at point x_0. The derivative of $f(x)$, which is the gradient at the point x_0, is defined by:

$$m = \underset{x \to 0}{Lim} \frac{f(x_0 + \Delta x) - f(x_0)}{\Delta x} \tag{12.2}$$

The information about the function contour or gradient can be obtained based on the correlation between the derivative and the slope of the tangent (Miller 2011). An optimization is defined as finding the optimal solution which maximizes or minimizes an objective function. If $f(x)$ is not strongly convex or concave, there will be several stationary points $(\bar{x})$.

In addition, there is a maximum or minimum point in a flat part, which indicates that the optimal solution is located at a stationary point. In fact, the optimal solution of an optimization problem is located at a point with a zero gradient. So, the optimal solution can be obtained by determination of the function contour. Accordingly, the derivative, which is based on the limit [Eq. (12.2)], is an important concept in optimization problems. Although many real-world problems have discrete variables, the limit concept just applies in continuous functions. Furthermore, for some continuous functions, the calculation of limit is so difficult because of the complexity of functions. In this case, derivative is calculated using numerical methods such as Newton interpolation, Lagrange and cubic spine (Patil and Verma 2006). The GE algorithm was developed using the following Taylor-series-based first- and second-order derivatives:

$$f'(x) = \frac{f(x + \Delta x) - f(x - \Delta x)}{2 \cdot \Delta x} \tag{12.3}$$

$$f''(x) = \frac{f(x + \Delta x) - 2f(x) + f(x - \Delta x)}{(\Delta x)^2} \tag{12.4}$$

where $f'(x)$ and $f''(x)$ = first and second order derivatives of $f(x)$, respectively (Kuo and Zulvia 2015).

12.2.2 Gradient-Based Algorithm

Optimization methods can be divided into two main groups: direct search and gradient-based search methods. Both groups start the search from a point and evaluate the other points in order to find an optimal solution till the stopping criterion is satisfied. In the direct search methods (region eliminating), the contour of a function is determined based on two or more points. Then, the search direction is limited to the search space which has a better initial point. The search space is decreased iteratively in different successful iterations in order to achieve the optimal solution. Golden search and Fibonacci search are two direct search methods (Bazaraa et al. 2013). The first- and second-order derivatives are applied in the function analysis using the gradient-based methods such as Newton–Raphson (Ypma 1995). The concept of this method applied in the GE algorithm is presented as follows.

In the Newton–Raphson method, the search process starts from an initial point and continues its movement to the point with a zero gradient. If the search is located at point x^t in iteration t, in next iteration it will be at point x^{t+1} which is located Δx^t from x^t. Since finding an extreme point is considered, the first derivative must be equal to zero. The Taylor series expansion is applied in order to estimate the first- and second-order derivatives [Eqs. (12.3) and (12.4)]. Furthermore, x^{t+1} is determined by:

$$x^{t+1} = x^t - \frac{\Delta x^t}{2} \cdot \frac{f(x^t + \Delta x^t) - f(x^t - \Delta x^t)}{f(x^t + \Delta x^t) - f(x^t) + f(x^t - \Delta x^t)} \tag{12.5}$$

The gradient-based methods apply numerical methods, instead of direct derivation, due to non-differentiability of the functions of optimization problems.

The global gradient of point $\overrightarrow{x_t}$ can be determined based on the sets of experimental candidates $\overrightarrow{t_i}$ $(i = 1, 2, \ldots, \lambda)$ in iteration t (Salomon 1998). In order to achieve $\overrightarrow{t_i}$, a random mutation, $\overrightarrow{z_i}$, is applied at current point $\overrightarrow{x_t}$. $\overrightarrow{e_t}$ includes all the information related to the experimental candidates and points in the gradient estimation direction.

$$\overrightarrow{e_t} = \frac{\overrightarrow{g_t}}{\|g_t\|} \tag{12.6}$$

$$\overrightarrow{g_t} = \sum_{i=1}^{\lambda} \left(f\left(\overrightarrow{\tau_i}\right) - f\left(\overrightarrow{x_t}\right) \right) \left(\overrightarrow{\tau_i} - \overrightarrow{x_t} \right), \tag{12.7}$$

where g_t = gradient direction in each iteration.

This method is improved in the evolutionary gradient search method (EGS). In this regard, the concept of central differencing is applied, instead of forward differencing (Salomon 1998). In the pseudo-code of GE, a simpler method is applied for gradient estimation (Wen et al. 2003). In the method, modification in direction from $\overrightarrow{x_i}$ to $\overrightarrow{x_j}$ in R^n, $\mathrm{dir}(i,j)$ is defined by:

$$\mathrm{dir}(i,j) = \begin{cases} 1 & if \quad x_{ik} \succ x_{jk} \\ 0 & if \quad x_{ik} = x_{jk} \quad \forall k = 1, 2, \ldots, n \\ -1 & if \quad x_{ik} \prec x_{jk} \end{cases} \tag{12.8}$$

Although the GE algorithm applied a central differencing approach-based gradient estimation method as well as EGS, the formulations of these two methods are completely different (Kuo and Zulvia 2015).

12.3   Mathematical Formulation of the GE Algorithm

As aforementioned, the GE algorithm applies a gradient-based method to determine the search direction. It is also based on population. This population includes a number of vectors which indicate the possible solutions. All vectors are updated by three operators: updating, jumping, and refreshing. Description of the application of the GE algorithm is presented for a minimization optimization problem in R^D as follows.

12.3.1   Solution Representation and Algorithm Initialization

In an optimization problem with D variables, the GE algorithm encodes the initial solution as a vector with D dimensions. Population P^t in iteration t is represented by $P^t = \left\{ X_i^t \,\middle|\, i = 1, \ldots, N \right\}$ which includes n vectors. $X_i^t = \left\{ x_{ij}^t \,\middle|\, \begin{array}{l} i = 1, \ldots, N \\ j = 1, \ldots, D \end{array} \right\}$ is the ith vector in iteration t which corresponds a feasible solution for the problem. In the initial step, it is necessary to determine six parameters including number of iterations T, number of vectors N, size of initial step λ, rate of jump J_r, rate of refreshing S_r, and rate of reduction ε.

Determining the number of iterations and vectors depends on the complexity of the problem. In complex problems, more iterations and more vectors are considered. J_r, S_r, and ε are in the interval [0, 1]. J_r is used when there is considerable modification in the vector direction. S_r and ε also manage vector regeneration and refreshing. Moreover, the value of ε is effective for acceleration of vector refreshing. It is necessary to determine the initial points for all vectors in the GE algorithm. In this regard, a simplest method is employed to generate random numbers.

12.3.2 *Vector Updating*

The updating rule, which controls the vector movement in order to reach a better position, includes two gradient-based and acceleration factor parts. The first part is the core of updating rule and is derived from the gradient-based methods, which are started from an arbitrary initial point and move gradually to the next point in a certain direction determined by the gradient.

Movement to the points with better values of objective function in the search direction is illustrated in Fig. 12.2. The GE algorithm also determines the search space which has better solutions. Due to the complexity of the problem, the gradient is not considered as the first-order derivative of the objective function and the GE algorithm applies central differencing instead. The search process of the GE algorithm is shown in Fig. 12.3.

As shown in Fig. 12.3a, the GE algorithm explores the search space to achieve a better area. If all the vectors in population move simultaneously to the same area, the search space becomes narrow. If there is a distraction direction for vector movement, the search process of the algorithm is performed in a wider range (Fig. 12.3b). The gradient-based updating rule applies the Newton–Raphson equation [Eq. (12.5)]. Since the main updating rule is individual-based, in order to

Fig. 12.2 Search direction for the original gradient-based method

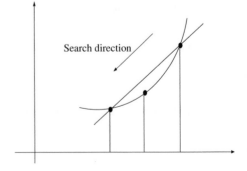

Search direction

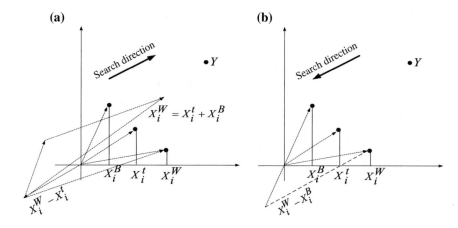

Fig. 12.3 Search direction for the GE algorithm

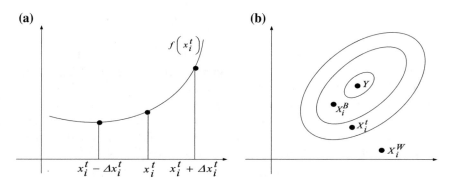

Fig. 12.4 Gradient approximation method modified from individual-based search to population-based search: **a** individual-based, **b** population-based

apply this method in the GE algorithm, which is population-based, some modifications are required. The updating rule in Eq. (12.5) includes the neighboring points of X^t which are $X^t - \Delta X$ and $X^t + \Delta X$. It is necessary to evaluate the neighboring points in order to reach the next point X^{t+1}. This evaluation is time consuming in population-based search for each vector. So, the GE algorithm replaces points $X^t - \Delta X$ and $X^t + \Delta X$ with the positions of two other vectors in population.

The transformation from the individual-based updating rule to the population-based updating rule is shown in Fig. 12.4. In individual-based search, X_i^t has two neighboring points $X_i^t + \Delta X_i^t$ and $X_i^t - \Delta X_i^t$. If $f(X_i^t)$ is a minimization problem, $f(X_i^t - \Delta X_i^t) \le f(X_i^t) \le f(X_i^t + \Delta X_i^t)$, which implies that point $X_i^t - \Delta X_i^t$ is better than point X_i^t and point $X_i^t + \Delta X_i^t$ is worse than point X_i^t. Since the GE algorithm is based on the population-based search method, there are many possible solutions in population. In fact, in addition to the worst and best vectors, there is a vector X_i^t which may have a neighboring with a worse or better position.

To update X_i^t in the GE algorithm, point $X_i^t - \Delta X_i^t$ is replaced with vector X_i^B [$X_i^B \in P^t$, $f(X_i^B) \leq f(X_i^t)$]   and $X_i^t + \Delta X_i^t$ is   replaced   with X_i^W [$X_i^W \in P^t$, $f(X_i^W) \leq f(X_i^t)$]. So, points $X_i^t - \Delta X_i^t$ and $X_i^t + \Delta X_i^t$ are substituted by vectors X_i^B and X_i^W, respectively. The GE algorithm also applies position X_i^t instead of the fitness of position $f(X_i^t)$ because applying a fitness value is time consuming. In order to expand the search, a random number $r_g \approx N(0,1)$ is added to the updating rule. Considering r_g ensures the distribution of vector movement in the GE algorithm. The updating rule in Eq. (12.5) can be transformed to GradientMove by:

$$Gradient\,Move = \left(r_g \cdot \frac{\Delta x_{ij}^t}{2}\right) \cdot \left(\frac{x_{ij}^W - x_{ij}^B}{x_{ij}^W - x_{ij}^t + x_{ij}^B}\right), \quad \forall j = 1,\ldots,D \qquad (12.9)$$

$$\Delta x_{ij}^t = \frac{\left|x_{ij}^t - x_{ij}^B\right| + \left|x_{ij}^W - x_{ij}^t\right|}{2}, \quad \forall j = 1,\ldots,D \qquad (12.10)$$

The acceleration factor Acc is used to accelerate the convergence of each vector. This process uses the best vector of a direction which is expressed by Eq. (12.11). In the GE algorithm, it is assumed that the best vector is the vector closest to the optimal solution. So, all other vectors will move in a better direction by considering the position of the best vector. Similar to the gradient updating rule, the acceleration factor in Eq. (12.11) is also multiplied by a random number $r_a \approx N(0,1)$, which ensures different sizes of steps for each vector (Kuo and Zulvia 2015).

$$Acc = r_a \cdot \left(y_i - x_{ij}^t\right), \quad \forall j = 1,\ldots,D, \qquad (12.11)$$

where $Y = \{y_i | j = 1,2,\ldots,D\}$, the best vector. Finally, the vector updating is conducted by (Kuo and Zulvia 2015):

$$u_{ij}^t = x_{ij}^t - GradientMove + Acc$$

$$u_{ij}^t = x_{ij}^t - \left(r_g \cdot \frac{\Delta x_{ij}^t}{2}\right) \cdot \left(\frac{x_{ij}^W - x_{ij}^B}{x_{ij}^W - x_{ij}^t + x_{ij}^B}\right) + r_a \left(y_i - x_{ij}^t\right) \quad \forall j = 1,\ldots,D \qquad (12.12)$$

where $u_{ij}^t \in U_i^t$ = transition vector which is obtained by updating X_i^t. Since vector X_i^t includes worse and better vectors, an additional process is necessary in order to determine the worst and best vectors in P^t because the main updating rule includes the neighboring vectors which have a worse or better fitness value. The worst and best vectors do not have any neighboring vector which is worse or better than themselves. In this regard, an additional process is required to determine the gradient of the objective function.

If W^t and B^t are the worst and best vectors of X_i^t, respectively, the values of x_{ij}^W and x_{ij}^B can be replaced with w_j and b_j, respectively by Eqs. (12.13)–(12.16) (Kuo and Zulvia 2015):

$$b_j = x_{ij}^t - \Delta x_{ij}^t \quad \forall j = 1, \ldots, D \tag{12.13}$$

$$\Delta x_{ij}^t = \frac{r + \left| x_{ij}^W - x_{ij}^t \right|}{2} \quad \forall j = 1, \ldots, D \tag{12.14}$$

$$w_j = x_{ij}^t + \Delta x_{ij}^t \quad \forall j = 1, \ldots, D \tag{12.15}$$

$$\Delta x_{ij}^t = \frac{\left| x_{ij}^t - x_{ij}^B \right| + \gamma}{2} \quad \forall j = 1, \ldots, D \tag{12.16}$$

where γ = size of the initial step which is predefined. γ can be a static or dynamic number. It can be decreased by increasing the number of iterations if it is a dynamic number. There are two ways for solving a maximization problem: (1) switching worse and better neighbors and (2) transforming into a minimization problem.

12.3.3   Vector Jumping

An appropriate search method must be able to explore the search space widely and deeply. The vector updating and vector jumping operators focus on deep and wide search, respectively. In the GE algorithm, the vector jumping operator is applied to avoid local optima. This operator just performs on a selective vector and modifies the movement direction. In the GE algorithm, J_r is considered for determining whether or not the vector must jump. If $r_{j \leq J_r} (r_j \approx N(0, 1))$, vector jumping to a transition vector U_i^t is given by (Kuo and Zulvia 2015):

$$u_{ij}^t = -u_{ij}^t + r_m \cdot \left(u_{ij}^t - x_{kj}^t \right) \quad \forall j = 1, \ldots, D, \tag{12.17}$$

where $x_{kj}^t \in X_{kj}^t$ in each random neighboring vector P^t, $\forall i \neq k$, and r_m = random number in the interval (0, 1). The process of vector jumping is schematically shown in Fig. 12.5.

Fig. 12.5 Vector jumping
operator

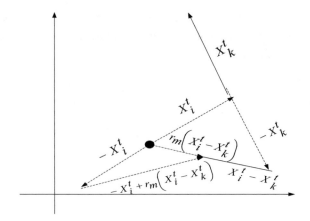

12.3.4   Vector Refreshing

The GE algorithm uses an elitist strategy. In iteration t, the transition vector U_i^t records the updated results and the jumped vector X_i^t. The next position of vector i, X_i^{t+1}, is replaced by U_i^t when the fitness value $f(U_i^t)$ is better than fitness value $f(x_i^t)$. Otherwise, it remains at X_i^t in next iteration (i.e., $X_i^{t+1} = X_i^t$). Using the elitist strategy, the GE algorithm ensures that each vector always moves to a better position. If the determination of a better position is difficult, problem arises. This situation takes place in the complex problems and the problems with many local optima. In this case, vector refreshing is performed for such a problematic vector. The GE algorithm records the position of vector X_i^t and the history of vector updating. The history of vector i which is $s_i \in [0, 1]$, provides information about the number of iterations that the vector cannot move to a better position. The newly generated vector s_i is set to one. When vector i is stuck in the same position, s_i is reduced by:

$$s_i = s_i - \varepsilon.s_i, \qquad (12.18)$$

where ε = rate of reduction. The predefined parameters S_r and ε are continuous numbers in the interval [0, 1]. They will be set when vector regeneration is required. If s_i is less than S_r, vector i must be regenerated (Kuo and Zulvia 2015). The flowchart of the GE algorithm is shown in Fig. 12.6. The parameters and variables used in the GE algorithm are presented in Table 12.1.

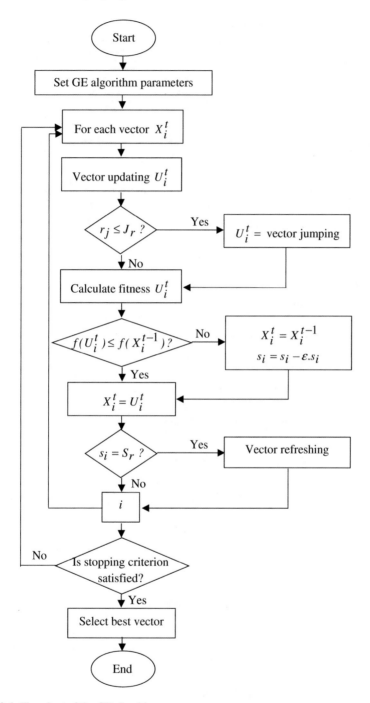

Fig. 12.6 Flowchart of the GE algorithm

Table 12.1 Characteristics of the GE algorithm

General algorithm	Gradient Evolution Algorithm
Decision variable	Vector which indicates the possible solutions
Solution	Movement to the point with a zero gradient
Old solution	Old vector of possible solutions
New solution	New vector of possible solutions
Best solution	Vector with the best fitness function
Fitness function	Gradient obtained by a gradient-based method
Initial solution	Randomly generated vector
Selection	–
Process of generating new solution	Vector updating, vector jumping, and vector refreshing

12.4   Pseudo-Code of GE

Begin

Step 1. Initialization

Set parameters including $T, N, \gamma, J_r, S_r, \varepsilon$

Generate an initial population P^0

Set an initial updating history $s_i = 1, \forall i = 1, ..., N$

Calculate the fitness value of each vector $f\left(X_i^0\right), \quad i = 1, ..., N$

Define the best vector $B^0 = X_i^0 \Leftrightarrow f\left(X_i^0\right) \leq f\left(X_j^0\right), \quad \forall j = 1, ..., N \, ; i \neq j$

Define the worst vector $W^0 = X_i^0 \Leftrightarrow f\left(X_i^0\right) \geq f\left(X_j^0\right), \quad \forall j = 1, ..., N \, ; i \neq j$

Record the best vector $Y = \{y_1, ..., y_D\}$

Step 2. For $t = 1 : T$

For $i = 1 : N$

Perform vector updating operator

If $X_i^{t-1} = B^{t-1}$

Update X_i^{t-1}

Else if $X_i^{t-1} = W^{t-1}$

Update X_i^{t-1}

Else

  Update X_i^{t-1}

End if

Perform vector jumping operator

  Generate r_j, $j = 1, ..., D$

Choose any neighborhood X_i^{t-1}

If $r_j \leq J_r$

  Jump a vector to transformation vector

End if

Calculate the fitness value $f\left(U_i^t\right)$

Perform vector refreshing operator

If $f\left(U_i^t\right) \prec f\left(X_i^{t-1}\right)$

  Set $X_i^t = U_i^t$

Else

  Set $X_i^t = X_i^{t-1}$ and update s_i

End if

  If $s_i \leq S_r$

    Regenerate X_i^t, calculate $f\left(X_i^t\right)$, and set $s_i = 1$

  End if

  Consider $i = i+1$

  End for i

  Update Y

  Repeat step 2, until the stopping criterion is satisfied

  End for t

  Step 3. Vector Y is the final solution

End

12.5   Conclusion

This chapter introduced the gradient evolution (GE) algorithm, a population-based meta-heuristic algorithm based on the concept of gradient. The gradient-based optimization methods evaluate the search space based on an individual-based approach. In order to apply these methods in the population-based algorithms, considerable modifications must be performed. The GE algorithm explores the search space based on a set of vectors in optimization problems and applies three main operators including vector updating, vector jumping, and vector refreshing. The vector updating operator based on the Tylor series expansion theorem transforms the updating rule for population-based search. The vector jumping operator avoids local optima and the vector refreshing operator is applied when a vector cannot move to another position in several iterations. This chapter detailed the underlying idea and the mathematical formulation of the GE algorithm.

References

Bazaraa, M. S., Sherali, H. D., & Shetty, C. M. (2013). *Nonlinear programming: Theory and algorithms* (3rd ed.). New Jersey, USA: Wiley.

Kuo, R. J., & Zulvia, F. E. (2015). The gradient evolution algorithm: A new metaheuristic. *Information Sciences, 316,* 246–265.

Kuo, R. J., & Zulvia, F. E. (2016, July 25–29). Cluster analysis using a gradient evolution-based k-means algorithm. In *IEEE congress on evolutionary computation (CEC).* Beijing, China: Peking University.

Larson, R., Hostetler, R. P., & Edwards, B. H. (2007). *Essential Calculus: Early Transcendental Functions* (1st ed.). New York, USA: Houghton Mifflin Company.

Miller, H. R. (2011). *Optimization: Foundations and applications* (1st ed.) New York, USA: Wiley.

Patil, P. B., & Verma, U. P. (2006). *Numerical computational methods* (1st ed.). Oxford, UK: Alpha Science International.

Salomon, R. (1998). Evolutionary algorithms and gradient search: Similarities and differences. *IEEE Transactions on Evolutionary Computation, 2*(2), 45–55.

Wen, J. Y., Wu, Q. H., Jiang, L., & Cheng, S. J. (2003). Pseudo-gradient based evolutionary programming. *Electronics Letters, 39*(7), 631–632.

Ypma, T. J. (1995). Historical development of the Newton-Raphson method. *SIAM Review, 37*(4), 531–551.

Chapter 13
Moth-Flame Optimization (MFO) Algorithm

Mahdi Bahrami, Omid Bozorg-Haddad and Xuefeng Chu

Abstract This chapter introduces the Moth-Flame Optimization (MFO) algorithm, along with its applications and variations. The basic steps of the algorithm are explained in detail and a flowchart is represented. In order to better understand the algorithm, a pseudocode of the MFO is also included.

13.1 Introduction

The MFO is a novel nature-inspired optimization algorithm based on the navigation method of moths through night by maintaining a fixed angle with respect to the Moon. Introduced by Mirjalili (2015) and tested on 29 benchmark functions and seven real engineering problems, MFO showed promising results in comparison to other nature-inspired algorithms. Since being introduced, MFO has been applied to real engineering problems in different fields of research.

Yamany et al. (2015) trained a multi-layer perceptron (MLP) using MFO. MFO-MLP was used to search for the weights and biases of the MLP to achieve the minimum error and the highest classification rate. Li et al. (2016a) proposed a new hybrid annual power load forecasting model based on the least squares support vector machine (LSSVM) and the MFO algorithm to forecast the annual power load essential for the planning, operation, and maintenance of an electric power system.

M. Bahrami · O. Bozorg-Haddad (✉)
Department of Irrigation and Reclamation Engineering, Faculty of Agricultural Engineering and Technology, College of Agriculture and Natural Resources, University of Tehran, 31587-77871 Karaj, Tehran, Iran
e-mail: OBHaddad@ut.ac.ir

M. Bahrami
e-mail: M.Bahrami9264@ut.ac.ir

X. Chu
Department of Civil and Environmental Engineering, North Dakota State University, Dept 2470, Fargo, ND 58108-6050, USA
e-mail: Xuefeng.Chu@ndsu.edu

© Springer Nature Singapore Pte Ltd. 2018
O. Bozorg-Haddad (ed.), *Advanced Optimization by Nature-Inspired Algorithms*, Studies in Computational Intelligence 720, DOI 10.1007/978-981-10-5221-7_13

The parameters for LSSVM were optimally determined using the MFO algorithm. Raju et al. (2016) used MFO for simultaneous optimization of secondary controller gains in a cascade controller proposed for automatic generation control of a two-area hydro-thermal system under a deregulated scenario. Zawbaa et al. (2016) proposed a feature selection algorithm based on MFO and applied it to machine learning for feature selection to find the optimal feature combination using the wrapper-based feature selection mode. MFO was exploited as a searching method to find the optimal feature set, maximizing classification performance. Ceylan (2016) used MFO to solve the harmonic elimination problem and minimize the total harmonic distortion. The simulation results showed that the MFO model solved the harmonic elimination problem and the total harmonic distortion minimization problem efficiently. Lal and Barisal (2016) applied MFO to evaluate the optimal gains of the fuzzy-based proportional, integral and derivative (PID) controllers in a microgird power generation system interconnected with a single area reheat thermal power system. Gope et al. (2016) used MFO to obtain an optimal bidding strategy of supplier considering double sided bidding under a congested power system. Jangir et al. (2016) solved five constrained benchmark functions of engineering problems using MFO and compared the results with other recognized optimization algorithms. MFO provided better results in various design problems in comparison to other optimization algorithms. Parmar et al. (2016) solved the optimal power flow (OPF) problem using MFO involving fuel cost reduction, active power loss minimization, and reactive power loss minimization. Comparing with other techniques such as flower pollination algorithm (FPA) and particle swarm optimization (PSO), MFO showed a better performance. Bentouati et al. (2016) applied MFO to solve the problem of OPF in the interconnected power system for the Algerian power system network with different objective functions. The results were compared with those obtained by artificial bee colony (ABC) and other metaheuristics. Allam et al. (2016) utilized MFO for the parameter extraction process of the three-diode model for the multi-crystalline solar cell\module, with the results being compared with those obtained by FPA and hybrid evolutionary (DEIM) algorithms. The results showed that MFO achieved the least root mean square error (RMSE), mean bias error (MBE), absolute error at the maximum power point (AEMPP), and best coefficient of determination. Buch et al. (2017) applied MFO to various nonconvex, nonlinear optimum power flow objective functions with five single objective functions. Comparing MFO with other stochastic methods showed that MFO obtained the optimum value with rapid and smooth convergence. Garg and Gupta (2017) used MFO to optimize the performance of open shortest path first (OSPF) algorithm which is the widely used, efficient algorithm to select the shortest path between the source and destination. The results for different scenarios showed the reduction in delay and energy consumption in the optimized OSPF compared to the traditional OSPF. Khalilpourazari and Pasandideh (2017) solved a multi-constrained Economic Order Quantity (EOQ) model using MFO and the interior-point method. To compare the results of the two methods, three measures, including objective function value, computation time, and the number of function evaluations were used. The results indicated that there was no significant difference between the average objective function values of MFO and the interior-point method, but MFO required significantly less computation time and fewer function evaluations.

Different versions of the MFO algorithm have been developed by other researchers in order to improve the performance of the algorithm in different fields of research. Bhesdadiya et al. (2016) developed a hybrid optimization algorithm based on PSO and MFO. PSO was used for the exploitation and MFO was utilized for the exploration phase. The proposed algorithm was tested on some unconstraint benchmark test functions along with some constrained/complex design problems and the obtained results demonstrated its effectiveness comparing to the standard PSO and MFO algorithms. Nanda (2016) modified the original MFO to handle multi-objective optimization problems. The proposed MOMFO used concepts such as the archive grid, coordinate based distance for sorting and non-dominance of solutions. In the tests of six benchmark mathematical functions, MOMFO achieved better accuracy and shorter computational time than non-dominated Sorting Genetic Algorithm-II (NSGA-II) and Multi-objective PSO (MOPSO). Muangkote et al. (2016) proposed an improved version of MFO for image segmentation to enhance the optimal multilevel thresholding of satellite images. The proposed multilevel thresholding moth-flame optimization algorithm (MTMFO) was tested for various satellite images. MTMFO provided more effective results with better accuracy than MFO.

Soliman et al. (2016) proposed two modified versions of MFO and used as a prediction tool for terrorist groups, and compared with the original MFO as well as ant lion optimizer (ALO), grey wolf optimization (GWO), PSO, and genetic algorithm (GA). The results proved that the modified versions of MFO achieved an advance over the original MFO algorithm. Li et al. (2016b) proposed an improved version of MFO based on the Lévy-flight strategy (LMFO) to improve the convergence and precision of MFO. The new LMFO algorithm increased the diversity of the population against premature convergence and made the algorithm jump out of local optimum more effectively. Compared with MFO and other heuristic methods, LMFO demonstrated its superior performance. Trivedi et al. (2016) applied MFO to economic load dispatch problems. They integrated MFO with Lévy flights to achieve the competitive results in case of both discrete and continuous control parameters.

13.2   Mapping the Navigation Method of Moths into Moth-Flame Optimization

MFO is a nature-inspired optimization algorithm based on the moths' navigation mechanism in the night, known as the transverse orientation. Moths can travel long distances in a straight line by maintaining a fixed angle with the Moon (Frank 2006).

Fig. 13.1 Moth's spiral
flying path around a light
source

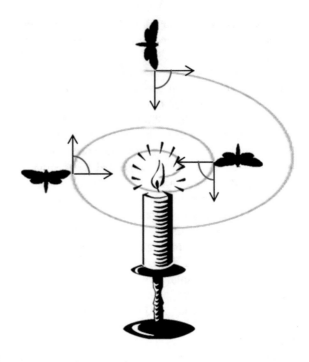

When encountered artificial lights, moths try to maintain a similar angle to the light
source and because of the close distance they get caught in a spiral path (Fig. 13.1).

The MFO assigns moths to different solutions in the solution space of the
optimization problem with each moth having its own fitness function value. Each
moth also has a flame which stores the best solution found by that moth. In each
iteration, the moths search the solution space by flying through a spiral path toward
their flames and update their positions.

MFO starts with the positions of moths randomly initialized within the solution
space. The fitness values of the moths are calculated which are the best individual
fitness values so far. The flame tags the best individual position for each moth. In
the next iteration, the moths' positions are updated based on a spiral movement
function toward their best individual positions tagged by a flame and the positions
of the flames are updated with new best individual positions. The MFO algorithm
continues updating the positions of the moths and flames and generating new
positions until the termination criteria are met. Table 13.1 lists the characteristics of
the MFO. Figure 13.2 shows the flowchart of the MFO.

Table 13.1 Characteristics of the MFO algorithm

General algorithm	Moth-flame optimization
Decision variable	Moths' positions in each dimension
Solution	Moths' positions
Old solution	Old positions of moths
New solution	New positions of moths
Best solution	Positions of flames
Fitness function	Distance between moth and flame
Initial solution	Random positions of moths
Selection	–
Process of generating new solution	Flying in a spiral path toward a flame

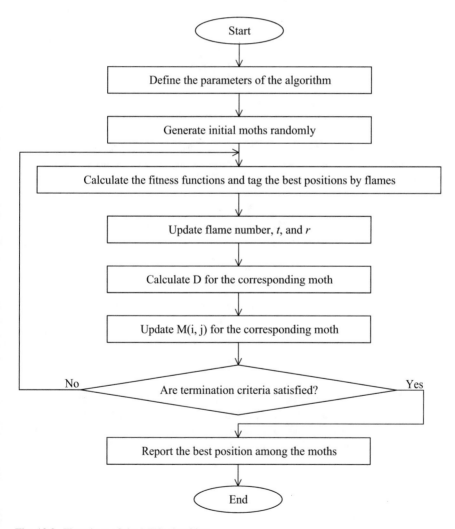

Fig. 13.2 Flowchart of the MFO algorithm

13.3   Creating the Initial Population of Moths

In MFO, each moth is assumed to have a position in a D-dimensional solution space. The set of moths can be expressed as (Mirjalili 2015):

$$
M = \begin{bmatrix}
m_{1,1} & m_{1,2} & \cdots & \cdots & m_{1,d} \\
m_{2,1} & m_{2,2} & \cdots & \cdots & m_{2,d} \\
. & . & . & . & . \\
. & . & . & . & . \\
m_{n,1} & m_{n,2} & \cdots & \cdots & m_{n,d}
\end{bmatrix}
\tag{13.1}
$$

in which n = number of moths; and d = number of dimensions in the solution space. The corresponding fitness function values for the moths are sorted in an array as follows (Mirjalili 2015):

$$
OM = \begin{bmatrix}
OM_1 \\
OM_2 \\
. \\
. \\
OM_n
\end{bmatrix}
\tag{13.2}
$$

Two other components of the MFO are the flame matrix representing the flames in the D-dimensional space and their corresponding fitness function vector, which can be respectively expressed as (Mirjalili 2015):

$$
F = \begin{bmatrix}
F_{1,1} & F_{1,2} & \cdots & \cdots & F_{1,d} \\
F_{2,1} & F_{2,2} & \cdots & \cdots & F_{2,d} \\
. & . & . & . & . \\
. & . & . & . & . \\
F_{n,1} & F_{n,2} & \cdots & \cdots & F_{n,d}
\end{bmatrix}
\tag{13.3}
$$

$$
OF = \begin{bmatrix}
OF_1 \\
OF_2 \\
. \\
. \\
OF_n
\end{bmatrix}
\tag{13.4}
$$

In MFO, moths and flames represent solutions, with moths searching the solution space in each iteration to find the optimal solution and the flames representing the best solution found by each moth. In other words, each moth searches the space around its flame and each time it finds a better solution. The position of the flame is then updated.

13.4   Updating the Moths' Positions

MFO uses three functions to initialize the random positions of the moths (I), move the moths in the solution space (P), and terminate the search operation (T):

$$\text{MFO} = (I, P, T) \tag{13.5}$$

Any random distribution can be used to initialize the moths' positions in the solution space. The implementation of the I function can be written as (Mirjalili 2015):

$$M(i,j) = (\text{ub}(i) - \text{lb}(j)) * \text{rand}() + \text{lb}(i) \tag{13.6}$$

in which ub and lb = arrays that respectively define the upper and lower bounds of variables.

The movement of moths in the solution space is based on the transverse orientation and modeled by using a logarithmic spiral subjected to the following conditions (Mirjalili 2015):

- Spiral's initial point should start from the moth
- Spiral's final point should be the position of the flame
- Fluctuation of the range of spiral should not exceed the search space.

Hence, the P function for the movement is defined as:

$$S(M_i, F_j) = D_i . e^{bt} . \cos(2\pi t) + F_j \tag{13.7}$$

in which b = a constant to define the shape of the logarithmic spiral; t = a random number between $[-1, 1]$; and D_i = distance between the i-th moth and the j-th flame, which is defined as (Mirjalili 2015):

$$D_i = \left| F_j - M_i \right| \tag{13.8}$$

The spiral movement of the moth around the flame guarantees the exploration and exploitation of the solution space. In order to prevent the moths getting trapped in local optima, the best solutions (flames) are sorted in each iteration and each moth flies around its corresponding flame based on the OF and OM matrices. In other words, the first moth flies around the best obtained solution, while the last moth flies around the worst obtained solution.

13.5   Updating the Number of Flames

In order to improve the exploitation of the MFO algorithm, Eq. (13.9) is used to decrease the number of the flames, and hence the moths only fly around the best solution in the final steps of the algorithm (Mirjalili 2015):

$$Flame\, no = round\left(N - l * \frac{N - l}{T} \right) \qquad (13.9)$$

in which l = current number of iterations; N = maximum number of flames; and T = maximum number of iterations. The decrease in the number of the flames balances the exploration and exploitation of the solution space.

13.6   Termination Criteria

The termination criterion determines when the algorithm is terminated. Selecting a good termination criterion has an important role to ensure a correct convergence of the algorithm. The number of iterations, the amount of improvement, and the running time are common termination criteria for the MFO.

13.7   Performance of the MFO

Mirjalili (2015) tested MFO on 29 benchmark functions and seven real engineering problems, and compared the results with those obtained by other well-known nature-inspired algorithms such as PSO, gravitational search algorithm (GSA), bat algorithm (BA), FPA, states of matter search (SMS), firefly algorithm (FA), and GA. MFO showed promising and competitive results for the benchmark test functions and the results of the real problems demonstrated the MFO's ability in dealing with challenging problems with constrained and unknown search spaces.

13.8 Pseudocode of the MFO

Begin

Input parameters of the algorithm and the initial data

Initialize the positions of moths and evaluate their fitness values

While (the stop criterion is not satisfied or $I < I_{max}$)

Update flame no.

OM = Fitness Function(M)

If iteration = 1

$F = \text{sort}(M)$

$OF = \text{sort}(OM)$

Else

$F = \text{sort}(Mt - 1, Mt)$

$OF = \text{sort}(Mt-1, Mt)$

End if

For $i = 1: N$

For $j = 1: D$

Update r and t

Calculate D with respect to the corresponding moth

Update $M(i,j)$ with respect to the corresponding moth

End for j

End for i

End While

Post-processing the results and visualization.

End

13.9   Conclusion

This chapter described moth-flame optimization (MFO), a new nature-inspired algorithm based on the transverse orientation method used by moths. MFO searches the decision space using a set of moths which report the fitness functions at each time step and tag the best solution by a flame. The movement of moths is based on the flames in a spiral path around their flames. In this chapter, a literature review of MFO was presented, showing the success of the algorithm for different optimization problems, along with different variations of the MFO algorithm developed by researchers. The flowchart and the pseudocode of MFO were also presented to help understand the detailed computational procedures of the algorithm.

References

Allam, D., Yousri, D. A., & Eteiba, M. B. (2016). Parameters extraction of the three diode model for the multi-crystalline solar cell/module using Moth-Flame Optimization Algorithm. *Energy Conversion and Management, 123,* 535–548.

Bentouati, B., Chaib, L., & Chettih, S. (2016). Optimal Power Flow using the Moth Flam Optimizer: A case study of the Algerian power system. *Indonesian Journal of Electrical Engineering and Computer Science, 1*(3), 431–445.

Bhesdadiya, R. H., Trivedi, I. N., Jangir, P., Kumar, A., Jangir, N., & Totlani, R. (2016, August 12–13). A novel hybrid approach particle swarm optimizer with moth flame optimizer algorithm. In *International Conference on Computer, Communication and Computational Sciences (ICCCCS), Advances in Intelligent Systems and Computing.* Ajmer, India.

Buch, H., Trivedi, I. N., & Jangir, P. (2017). Moth flame optimization to solve optimal power flow with non-parametric statistical evaluation validation. *Cogent Engineering, 4*(1).

Ceylan, O. (2016, November 3–5). Harmonic elimination of multilevel inverters by moth-flame optimization algorithm. In *International Symposium on Industrial Electronics (INDEL).* Republic of Srpska, Bosnia and Herzegovina: IEEE.

Frank, K. D. (2006). Effects of artificial night lighting on moths. In C. Rich & T. Longcore (Eds.), *Ecological consequences of artificial night lighting* (pp. 305–344). Washington, DC: Island Press.

Garg, P., & Gupta, A. (2017). Optimized open shortest path first algorithm based on moth flame optimization. *Indian Journal of Science and Technology, 9*(48).

Gope, S., Dawn, S., Goswami, A. K., & Tiwari, P. K. (2016, November 22–25). Moth Flame Optimization based optimal bidding strategy under transmission congestion in deregulated power market. In *Region 10 Conference (TENCON).* Marina Bay Sands, Singapore: IEEE.

Jangir, N., Pandya, M. H., Trivedi, I. N., Bhesdadiya, R. H., Jangir, P., & Kumar, A. (2016, March 5–6). Moth-Flame Optimization algorithm for solving real challenging constrained engineering optimization problems. In *Students' Conference on Electrical, Electronics and Computer Science (SCEECS).* Bhopal, India: IEEE.

Khalilpourazari, S., & Pasandideh, S. H. R. (2017). Multi-item EOQ model with nonlinear unit holding cost and partial backordering: Moth-flame optimization algorithm. *Journal of Industrial and Production Engineering, 34*(1), 42–51.

Lal, D. K., & Barisal, A. K. (2016, December 27–28). Load frequency control of AC microgrid interconnected thermal power system. In *International Conference on Advanced Material Technologies (ICAMT).* Andhra Pradesh, India.

Li, C., Li, S., & Liu, Y. (2016a). A least squares support vector machine model optimized by moth-flame optimization algorithm for annual power load forecasting. *Applied Intelligence, 45* (4), 1166–1178.

Li, Z., Zhou, Y., Zhang, S., & Song, J. (2016b). Lévy-flight moth-flame algorithm for function optimization and engineering design problems. *Mathematical Problems in Engineering.* doi:10.1155/2016/1423930.

Mirjalili, S. (2015). Moth-flame optimization algorithm: A novel nature-inspired heuristic paradigm. *Knowledge-Based Systems, 89,* 228–249.

Muangkote, N., Sunat, K., & Chiewchanwattana, S. (2016, July 13–15). Multilevel thresholding for satellite image segmentation with moth-flame based optimization. In *The 13th International Joint Conference on Computer Science and Software Engineering.* Khon Kaen, Thailand.

Nanda, S. J. (2016, September 21–24). Multi-objective Moth Flame Optimization. In *Advances in Computing, Communications and Informatics (ICACCI).* Jaipur, India: IEEE.

Parmar, S. A., Pandya, M. H., Bhoye, M., Trivedi, I. N., Jangir, P., & Ladumor, D. (2016, April 7–8). Optimal active and Reactive Power dispatch problem solution using Moth-Flame Optimizer algorithm. In *International Conference on Energy Efficient Technologies for Sustainability (ICEETS).* Nagercoil, India: IEEE.

Raju, M., Saikia, L. C., & Saha, D. (2016, November 22–25). Automatic generation control in competitive market conditions with moth-flame optimization based cascade controller. In *Region 10 Conference (TENCON).* Marina Bay Sands, Singapore: IEEE.

Soliman, G. M. A., Khorshid, M. M. H., & Abou-El-Enien, T. H. M. (2016, July). Modified moth-flame optimization algorithms for terrorism prediction. *International Journal of Application or Innovation in Engineering and Management, 5,* 47–58.

Trivedi, I. N., Kumar, A., Ranpariya, A. H., & Jangir, P. (2016, April 7–8). Economic Load Dispatch problem with ramp rate limits and prohibited operating zones solve using Levy Flight Moth-Flame optimizer. In *International Conference on Energy Efficient Technologies for Sustainability (ICEETS).* Nagercoil, India.

Yamany, W., Fawzy, M., Tharwat, A., & Hassanien, A. E. (2015, December 29–30). Moth-flame optimization for training multi-layer perceptrons. In *11th International Computer Engineering Conference (ICENCO).* Giza, Egypt: IEEE.

Zawbaa, H. M., Emary, E., Parv, B., & Sharawi, M. (2016, July 24–29). Feature selection approach based on moth-flame optimization algorithm. In *Evolutionary Computation (CEC).* IEEE.

Chapter 14
Crow Search Algorithm (CSA)

Babak Zolghadr-Asli⑩**, Omid Bozorg-Haddad and Xuefeng Chu**

Abstract The crow search algorithm (CSA) is novel metaheuristic optimization algorithm, which is based on simulating the intelligent behavior of crow flocks. This algorithm was introduced by Askarzadeh (2016) and the preliminary results illustrated its potential to solve numerous complex engineering-related optimization problems. In this chapter, the natural process behind a standard CSA is described at length.

14.1 Introduction

In the last several decades, optimization played a crucial role in many aspects of various problems, including but not limited to engineering problems. Often, such problems include complicated objective functions, numerous decision variables, and a considerable number of constraints, which adds complexity to an already complicated optimization problem. The aforementioned characteristics limit the efficiency of traditional optimization techniques. Consequently, the search for an alternative method leads to a new field of study—swarm intelligence (SI), which was introduced by Beni and Wang in the late 1980s (Bei and Wang 1993). SI, ultimately, aims to imitate the social intelligence of the nature's group living creatures (Bonabeau et al. 1999). Each newly proposed algorithm attempts to

B. Zolghadr-Asli · O. Bozorg-Haddad (✉)
Department of Irrigation and Reclamation Engineering, Faculty of Agricultural Engineering and Technology, College of Agriculture and Natural Resources, University of Tehran, 3158777871 Karaj, Iran
e-mail: OBHaddad@ut.ac.ir

B. Zolghadr-Asli
e-mail: ZolghadrBabak@ut.ac.ir

X. Chu
Department of Civil and Environmental Engineering, North Dakota State University, Dept 2470, Fargo, ND 58108-6050, USA
e-mail: Xuefeng.Chu@ndsu.edu

© Springer Nature Singapore Pte Ltd. 2018
O. Bozorg-Haddad (ed.), *Advanced Optimization by Nature-Inspired Algorithms*,
Studies in Computational Intelligence 720, DOI 10.1007/978-981-10-5221-7_14

improve two main features: (1) decreasing the distance between the reported solutions and the actual global optima; and/or (2) reducing the solution searching time. Although each proposed optimization algorithm has its unique characteristics, with both merits and drawbacks, it has been proven that there is no single algorithm that could outperform all its rivals (Wolpert and Macready 1997). Subsequently, a wide range of alternative novel optimization algorithms have been proposed, each of which has its exclusive advantages.

One of these newly proposed algorithms is the crow search algorithm (CSA), which was initially introduced by Askarzadeh (2016). CSA attempts to imitate the social intelligence of crow flock and their food gathering process. The primary results illustrated the improved efficiency of CSA over many conventional optimization algorithms, such as genetic algorithm (GA), particle swarm optimization (PSO), and harmony search (HS), in both convergence time and the accuracy of the results (Askarzadeh 2016). Ultimately, it can be concluded that CSA is a proper alternative method for solving complex engineering optimization problems.

14.2   Crow Flock's Food Gathering Imitation

Crows are a widely distributed genus of birds, which have been credited with intelligence throughout folklore. Recent experiments investigating the cognitive abilities of crows have begun to reveal the intelligence capability of these species (Emery and Clayton 2004, 2005; Prior et al. 2008). The studies have demonstrated that some species of crows are not only superior in intelligence to other birds but also rival many nonhuman primates. Observations of the crows' tool use in the wild are an example of their complex cognition (Emery and Clayton 2004). Further studies have also revealed their self-awareness, face recognition capabilities, sophisticated communication techniques, and food storing and retrieving skills across seasons (Emery and Clayton 2005; Prior et al. 2008).

Interestingly, a crow individual has a tendency to tap into the food resources of other species, including the other crow members of the flock. In fact, each crow attempts to hide its excess food in a hideout spot and retrieve the stored food in the time of need. However, the other members of the flock, which have their own food reservation spots as well, try to tail one another to find these hiding spots and plunder the stored food. Nevertheless, if a crow senses that it has been pursuited by other members of the flock, in order to lose the tail and deceive the plunderer, it maneuvers its path into a fallacious hideout spot (Clayton and Emery 2005). Plainly, the aforementioned is the core principles of the CSA, in which each crow individual searches the decision space for hideouts with the best food resources (i.e., the global optima from the point of view of optimization). Thus, each crow individual's motion is induced by two main features: (1) finding the hideout spots of the other members of the flock; and (2) protecting its own hideout spots.

In the standard CSA, the flock of crows spread and search throughout the decision space for the perfect hideout spots (global optima). Since any efficient optimization algorithm should be compatible with arbitrary dimensions and each arbitrary dimension is to represent a decision variable, a d-dimensional environment is assumed for the search space. Initially, it is assumed that N crow individuals (the flock size) occupy a position in the d-dimensional space, randomly. The position of the ith crow individual at the tth iteration in the search space is represented by $x_{(i,t)}$, which is, in fact, a feasible array of the decision variables. Additionally, each crow individual can memorize the location of the best encountered hideout spot. At the tth iteration, the position of the hideout spot of the ith crow individual is represented by $m_{(i,t)}$, which is the best position that the ith crow individual has spotted, so far.

Subsequently, each crow individual shall make a motion based on the two basic principles of the CSA: (1) protecting its own hideout spot; and (2) detecting the other members' hideout spots. Assume that at the tth iteration, the jth crow individual attempts to retrieve food from its hideout spot $[m_{(j,t)}]$, while the ith crow decides to tail the jth crow individual, in order to plunder its stored food. In such circumstances, two situations may occur: (1) the jth crow individual could not detect that it has been tailed leading to the reveal of the hideout spot to the ith crow individual; or (2) the jth crow individual senses the presence of a plunderer, which leads to a deceiving maneuver by the jth crow.

In the first case, the lack of attention of the jth crow individual would enable the ith crow to spot and plunder the jth crow's hideout spot. In such a case, the repositioning of the ith crow can be obtained as follows (Askarzadeh 2016):

$$x_{(i,t+1)} = x_{(i,t)} + r_i \times fl_{(i,t)} \times \left[m_{(j,t)} - x_{(i,t)} \right] \tag{14.1}$$

in which r_i = a random number with the uniform distribution and the range of $[0,1]$; and $fl_{(i,t)}$ = flight length of the ith crow individual at the tth iteration.

It is worth to be mentioned that $fl_{(i,t)}$ is one of the algorithm's parameters and it can affect the searching capability of the algorithm. Assume that smaller values of fl lead to the local search at the vicinity of $x_{(i,t)}$, while larger values of fl would widen the searching space. In terms of optimization, smaller values of fl would help intensify the results, while larger values of fl would diversify the results. Both well-intensification and -diversification are the characteristics of an efficient optimization algorithm (Gandomi et al. 2013).

There could also be the case, in which the jth crow individual would sense that it had been tailed by one of the members of the flock (say the ith crow). As a result, in order to protect its food supply from the plunderer, the jth crow would deceitfully fly over a non-hideout spot. To imitate such an action in the CSA, a random place in the d-dimensional decision space would be assumed for the ith crow.

In summary, the tailing motion of crow individuals for the aforementioned two cases can be expressed as (Askarzadeh 2016)

$$x_{(i,t+1)} = \begin{cases} x_{(i,t)} + r_i \times fl_{(i,t)} \times \left[m_{(j,t)} - x_{(i,t)} \right] & r_j \geq AP_{(j,t)} \\ \text{a random position} & \text{otherwise} \end{cases} \tag{14.2}$$

in which r_j = a random number with the uniform distribution and the range of $[0,1]$; and $AP_{(j,t)}$ = the awareness probability of the jth crow at the tth iteration.

As mentioned previously, an efficient metaheuristic algorithm should provide a good balance between diversification and intensification (Yang 2010). In the CSA, intensification and diversification are mainly controlled by two parameters: the flight length (fl) and the awareness probability (AP). By decreasing the awareness probability, the chance of detecting the hideout spots by the members of the crow flock would increase. As a result, CSA tends to focus the search on the vicinity of the hideout spots. Thus, it can be assumed that smaller values of AP would amplify the intensification aspect of the algorithm. On the other hand, by increasing the awareness probability, the flock of crows is more likely to search the decision space in a random manner for, in fact, such an action would decrease the chance of discovering the real hideout spots by the plunderers. As a result, larger values of AP would amplify the diversification aspect of the algorithm.

14.3 CSA Implementation for Optimization

For an efficient implementation of a metaheuristic algorithm, one needs to tune the parameters of the algorithm. Parameter setting, however, is a time-consuming process. Thus, the algorithms with a limited number of parameters are easier to be implemented in different optimization problems. The aforementioned addresses one of the major advantages of the CSA over many conventional metaheuristic algorithms, since it has only two major parameters that require tuning (i.e., fl and AP). After the parameter adjustment, the flock size (N) and the maximum number of iterations (T) are assumed, as well.

The first step is to locate N crows, randomly, in a d-dimensional decision space. Since the crows have no experiences at the initial iteration, it is assumed that they have hidden their foods at their initial positions. After the initial step, the CSA relocates each crow individual, say the ith crow, as follows: the ith crow individual would assume the role of a plunderer for a randomly selected member of the flock, say the jth crow. Using Eq. (14.2), the new position of the ith crow is calculated.

To avoid unfeasible answers, it is suggested in the standard CSA to check the feasibility of the new location in the decision space. If an unfeasible location is generated in the latter process, the crow must stay still. An alternative for such a procedure is the implementation of a penalty function for unfeasible answers. In any case, the crow updates its memory as follows (Askarzadeh 2016):

$$m_{(i,t)} = \begin{cases} x_{(i,t+1)} \text{ if } f\left[x_{(i,t+1)}\right] & \text{is better than } f\left[m_{(i,t)}\right] \\ m_{(i,t)} & \text{otherwise} \end{cases} \qquad (14.3)$$

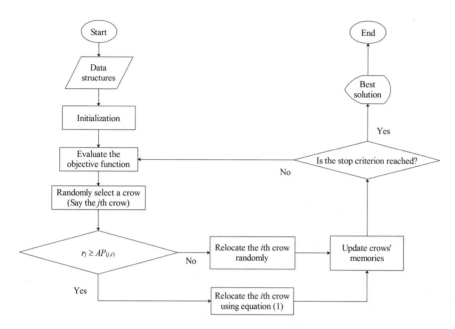

Fig. 14.1 Flowchart of the standard CSA

Table 14.1 Characteristics of the CSA

General algorithm	Crow search algorithm
Decision variable	Crow's position in each dimension
Solution	Crow's position
Old solution	Old position of a crow
New solution	New position of a crow
Best solution	Any crow with the best fitness
Fitness function	The value of discovered hideout spots
Initial solution	Randomly selected position of a crow
Selection	–
Process of generating new solution	Locating the flock members' hideout spots

in which $f[]$ = objective function. These steps are repeated until the termination criterion is satisfied. At that point, the best position that is memorized by the members of the crow flock is reported as the optimum solution. Figure 14.1 illustrates the flowchart of the standard CSA. Additionally, Table 14.1 summarizes the characteristics of the CSA.

14.4   Pseudo Code of the CSA

Begin

Define population size (N), the maximum number of iterations (T), flight length (fl), and awareness probability (AP).

Randomly initialize the positions of the flock of crows.

Initialize and record the crows' memories.

Set the iteration counter t=1

While (the termination criterion is not satisfied)

    For i=1:N

        Randomly choose one of the crows to follow (jth crow)

        Generate r_j

        If $r_j \geq AP_{(j,t)}$

$$x_{(i,t+1)} = x_{(i,t)} + r_i \times fl_{(i,t)} \times [m_{(j,t)} - x_{(i,t)}]$$

        Else

$$x_{(i,t+1)} = \text{a random place in the search space}$$

        End if

    End for i

    Check the feasibility of the new positions.

    Evaluate the new positions of the crows.

    Update the memories of crow individuals.

End while

Post-process and visualize the results.

End

14.5   Conclusion

This chapter described the crow search algorithm (CSA), which is a novel, yet relatively new metaheuristic optimization algorithm, based on the intelligent behavior of crows. CSA is a population-based optimization algorithm, with mainly two adjustable parameters (flight length and awareness probability). Such characteristics make CSA a viable option for complex engineering optimization problems. In the final section, a pseudocode of the standard CSA was also presented.

References

Askarzadeh, A. (2016). A novel metaheuristic method for solving constrained engineering optimization problems: Crow search algorithm. *Computers & Structures, 169,* 1–12.

Beni, G., & Wang, J. (1993). Swarm intelligence in cellular robotic systems. In P. Dario, G. Sandini, & P. Aebischer (Eds.), *Robots and Biological Systems: Towards a New Bionics?* Berlin, New York, NY: Springer.

Bonabeau, E., Dorigo, M., & Theraulaz, G. (1999). *Swarm intelligence: From natural to artificial systems.* New York, NY: Oxford University Press.

Clayton, N., & Emery, N. (2005). Corvid cognition. *Current Biology, 15*(3), R80–R81.

Emery, N. J., & Clayton, N. S. (2004). The mentality of crows: Convergent evolution of intelligence in corvids and apes. *Science, 306*(5703), 1903–1907.

Emery, N. J., & Clayton, N. S. (2005). Evolution of the avian brain and intelligence. *Current Biology, 15*(23), R946–R950.

Gandomi, A. H., Yang, X. S., & Alavi, A. H. (2013). Cuckoo search algorithm: A metaheuristic approach to solve structural optimization problems. *Engineering with Computers, 29*(1), 17–35.

Prior, H., Schwarz, A., & Güntürkün, O. (2008). Mirror-induced behavior in the magpie (Pica pica): Evidence of self-recognition. *PLoS Biology, 6*(8), e202.

Wolpert, D. H., & Macready, W. G. (1997). No free lunch theorems for optimization. *IEEE Transactions on Evolutionary Computation, 1*(1), 67–82.

Yang, X. S. (2010). *Nature-inspired metaheuristic algorithms.* Frome, UK: Luniver press.

Chapter 15
Dragonfly Algorithm (DA)

Babak Zolghadr-Asli⬮, Omid Bozorg-Haddad and Xuefeng Chu

Abstract The dragonfly algorithm (DA) is a new metaheuristic optimization algorithm, which is based on simulating the swarming behavior of dragonfly individuals. This algorithm was developed by Mirjalili (2016) and the preliminary studies illustrated its potential in solving numerous benchmark optimization problems and complex computational fluid dynamics (CFD) optimization problems. In this chapter, the natural process behind a standard DA is described at length.

15.1 Introduction

In the past decades, the natural swarming behavior of species has been the source of inspiration for a wide range of metaheuristic optimization algorithms. In fact, the aforementioned is the fundamental basis of swarm intelligence (SI), which was first proposed by Beni and Wang in the late 1980s (Beni and Wang 1993). SI, ultimately, aims to simulate the collective and social intelligence of nature's group living creatures (Bonabeau et al. 1999). Although both SI and traditional evolutionary algorithms (EAs), such as genetic algorithm (GA), have undeniable

B. Zolghadr-Asli · O. Bozorg-Haddad (✉)
Department of Irrigation and Reclamation Engineering, Faculty of Agricultural Engineering and Technology, College of Agriculture and Natural Resources, University of Tehran, 3158777871 Karaj, Iran
e-mail: OBHaddad@ut.ac.ir

B. Zolghadr-Asli
e-mail: ZolghadrBabak@ut.ac.ir

X. Chu
Department of Civil and Environmental Engineering, North Dakota State University, Dept 2470, Fargo, ND 58108-6050, USA
e-mail: Xuefeng.Chu@ndsu.edu

© Springer Nature Singapore Pte Ltd. 2018
O. Bozorg-Haddad (ed.), *Advanced Optimization by Nature-Inspired Algorithms*,
Studies in Computational Intelligence 720, DOI 10.1007/978-981-10-5221-7_15

advantages over traditional optimization methods in dealing with complex optimization problems, some may prefer the SI-based algorithms over EAs. First, there are fewer controlling parameters in the SI-based algorithms. Second, the SI-based algorithms are equipped with fewer operators than most traditional EAs (Mirjalili 2016).

The basic principles of the SI-based algorithms are centered around an iterative process, in which the SI searches through the decision space for arrays of decision variables, resulting in an optimum solution. This nature-inspired process is intended to imitate a natural feature that has been evolved over millions of years (Gandomi and Alavi 2012). Since the introduction of the SI-based algorithms, many promising metaheuristic algorithms have been proposed. These algorithms intend to increase the pace and accuracy of the searching process of the decision space, making them more suitable to solve complicated optimization problems.

Dragonfly algorithm (DA) is a newly proposed SI-based optimization algorithm. Dragonflies are majestic creatures. According to the discovered fossils, they may have evolved for more than 300 million years. Accordingly, there are up to 3000 different species of this insect around the world (Thorp and Rogers 2014). Dragonflies are considered as small carnivorous predators, eating a wide variety of insects ranging from small midges and mosquitoes to butterflies, moths, and damselflies. Although dragonflies are swift and agile fliers, some predators, such as swallows, are fast enough to hunt them as preys. A fascinating fact about dragonflies is perhaps their unique swarming behavior. They may swarm only for hunting or migration purposes (Mirjalili 2016). The former is referred to as the static (feeding) swarm (SS), and the latter is known as the dynamic (migratory) swarm (DS).

In the static swarm, dragonflies form small groups and maneuver over small areas. In this swarming behavior, which can be characterized by the local movements and swift changes in the flying orientation, each dragonfly individual tends to hunt flying preys. In a dynamic swarm, however, a massive number of dragonfly individuals form a migrating swarm, which would travel in one direction, and over long distances (Russell et al. 1998; Wikelski et al. 2006).

The aforementioned swarming behaviors of dragonflies are the main source of inspiration for the DA, for in fact, these behaviors are in line with the two main characteristics of a metaheuristic optimization algorithm: intensification (also known as exploitation) and diversification (also known as exploration) (Gandomi et al. 2013). The static swarming behavior enables dragonflies to create sub-swarms and investigate the presence of promising optima in numerous, yet small areas of the decision space (diversification). In a dynamic swarm, however, dragonflies will migrate in massive swarms toward what are the most promising locations for the global optimum, which in other words is a description of the intensification phase of an optimization algorithm. These are, in fact, the basic mechanism behind the DA.

The DA was initially proposed by Mirjalili (2016), and the preliminary studies have demonstrated its potential to outperform existing algorithms in solving both benchmark test problems and complicated engineering problems of computational fluid dynamics (CFD). The DA was also modified to better deal with binary [binary dragonfly algorithm (BDA)], and multi-objective optimization problems [multi-objective dragonfly algorithm (MODA)] (Mirjalili 2016). As a compatible and efficient algorithm, the DA can be an already promising alternative for solving complex engineering optimization problems. The following sections will focus on the basic principles of a standard DA.

15.2   Dragonflies' Swarming Patterns

The behavior of swarms follows three primitive principals (Reynolds 1987):

- Separation: The static collision avoidance of the dragonfly individuals in a neighborhood.
- Alignment: The velocity matching of dragonfly individuals in a neighborhood.
- Cohesion: The tendency of dragonfly individuals toward a neighborhood's center of the mass.

Additionally, any swarms of living creatures would also follow their survival instincts. Hence, all of the dragonfly individuals should be attracted toward food sources (food attraction) and distracted outward predators (predator distraction). In result, the swarming behavior of the dragonfly community can be explained by these five main factors (Fig. 15.1).

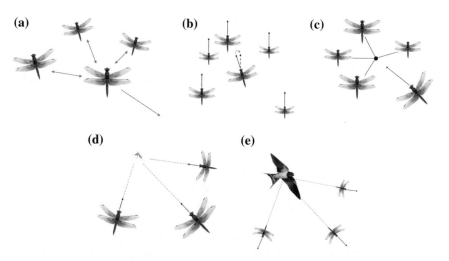

Fig. 15.1 Primitive corrective patterns of dragonfly individuals in a swarm: **a** Separation; **b** Alignment; **c** Cohesion; **d** Food Attraction; and **e** Predator distraction

In order to simulate the swarming behavior of dragonflies, the aforementioned characteristics must be mathematically modeled as follows:

The separation motion can be expressed as Reynolds (1987):

$$S_{(i,t)} = -\sum_{j=1}^{N} X_{(i,t)} - X_{(j,t)} \qquad (15.1)$$

in which $X_{(i,t)}$ = position of the ith dragonfly individual in the tth iteration; $X_{(j,t)}$ = position of the jth neighboring dragonfly individual in the tth iteration; N = number of neighboring dragonfly individuals; and $S_{(i,t)}$ = separation motion for the ith dragonfly individual in the tth iteration.

The alignment motion is calculated by Mirjalili (2016):

$$A_{(i,t)} = \frac{\sum_{j=1}^{N} V_{(j,t)}}{N} \qquad (15.2)$$

in which $V_{(j,t)}$ = velocity of the jth neighboring dragonfly individual in the tth iteration; and $A_{(i,t)}$ = alignment motion for the ith dragonfly individual in the tth iteration.

The cohesion motion can be measured by Mirjalili (2016):

$$C_{(i,t)} = \frac{\sum_{j=1}^{N} X_{(j,t)}}{N} - X_{(i,t)} \qquad (15.3)$$

in which $C_{(i,t)}$ = cohesion motion for the ith dragonfly individual in the tth iteration.

The food attraction motion is calculated by Mirjalili (2016):

$$F_{(i,t)} = X_{(\text{food},t)} - X_{(i,t)} \qquad (15.4)$$

in which $X_{(\text{food},t)}$ = position of the food source in the tth iteration; and $F_{(i,t)}$ = food attraction motion for the ith dragonfly individual in the tth iteration. The food is considered as the dragonfly individual with the best objective function observed so far.

The predator distraction is quantified by Mirjalili (2016):

$$E_{(i,t)} = X_{(\text{enemy},t)} + X_{(i,t)} \qquad (15.5)$$

in which $X_{(\text{enemy},t)}$ = position of the predator in the tth iteration; and $E_{(i,t)}$ = predator distraction motion for the ith dragonfly individual in the tth iteration. The predator is considered as the dragonfly individual with the worst objective function observed so far.

The combination of the aforementioned motions can predict the corrective pattern of dragonfly individuals in each iteration. The positions of dragonflies individuals are updated in each iteration using the current position of the dragonfly individual $[X_{(i,t)}]$ and a step vector $[\Delta X_{(i,t)}]$. In fact, the introduced step vector is analogous to the velocity vector in the particle swamps optimization (PSO) algorithm, and the procedure for updating the positions of dragonfly individuals in the DA is based on the framework of the PSO algorithm. The step vector, which demonstrates the motion orientation for each dragonfly individual, is defined as Mirjalili (2016):

$$\Delta X_{(i,t+1)} = \left(s \times S_{(i,t)} + a \times A_{(i,t)} + c \times C_{(i,t)} + f \times F_{(i,t)} + e \times E_{(i,t)}\right) + w \times \Delta X_{(i,t)} \tag{15.6}$$

in which s = separation weight; a = alignment weight; c = cohesion weight; f = food attraction weight; e = predator distraction weight; and w = inertia weight. After calculating the step vector, the updated position vectors are calculated by:

$$X_{(i,t+1)} = X_{(i,t)} + \Delta X_{(i,t)} \tag{15.7}$$

By tampering the separation, alignment, cohesion, food attraction, and predator weights (s, a, c, f, e, and w), different diversification and intensification behaviors can be achieved by the optimization. The aforementioned weights are, in fact, the DA's parameters and should be modified for each set of optimization problems in order to achieve preferable results. Additionally, neighboring of dragonfly individuals is crucial to the performance of the DA. In fact, an improper neighboring detecting mechanism could interfere with the convergence of the DA's results. This is due to the fact that in each algorithm, to detect the potential location of global optima, the searching mechanism should initially investigate the entire decision space thoroughly. This phase, which is better known as the desperation, requires the dragonfly individuals to spread through the search space. Yet, to locate the global optima, these individual dragonflies are required to converge and move toward the plausible locations of the global optima. This phase is known as the intensification. As discussed earlier, the swarming behavior of dragonflies can be categorized into two general motions: static swarming and dynamic swarming. Dragonflies tend to align their flying while maintaining proper separation and cohesion in a dynamic swarming (intensification). In a static swarm, however, alignments are very low while cohesion is high to attract prey. Therefore, we assign dragonflies with high alignment and low cohesion weights when exploring the search space. For transcending between diversification to intensification, the radii of neighborhoods are to increase proportionally to the number of iterations. This way, as the optimization proceeds, more dragonfly individuals can induce the motion of one another, causing the swarm to converge toward the possible location of the global optimum. Another way to balance diversification and intensification is to adaptively tune the DA's parameters (i.e., s, a, c, f, e, and w) during the optimization.

However, in order to increase the odds of investigating the entire decision space by any optimization algorithm, a random motion needs to be introduced to the searching mechanism. As a result, to improve the randomness, stochastic behavior, and exploration of the dragonfly individuals, they are required to fly around the search space using random walk (Lévy flight) when no neighboring solutions in the vicinity are detected. In this case, the positions of dragonflies are updated by Mirjalili (2016):

$$X_{(i,t+1)} = X_{(i,t)} + \text{Lévy}(d) \times X_{(i,t)} \tag{15.8}$$

in which d = number of decision variables; and Lévy(d) = Lévy flight function that is given by Yang (2010):

$$\text{Lévy}(d) = 0.01 \times \frac{r_1 \times \sigma}{|r_2|^{\frac{1}{\beta}}} \tag{15.9}$$

in which r_1 and r_2 = two random numbers in the range of [0,1]; β = constant, which is equal to 1.5 according to the measured values of the dragonfly individuals' movement (Mirjalili 2016); and σ is calculated by Yang (2010):

$$\sigma = \left(\frac{\Gamma(1+\beta) \times \sin\left(\frac{\pi\beta}{2}\right)}{\Gamma\left(\frac{1+\beta}{2}\right) \times \beta \times 2^{\left(\frac{\beta-1}{2}\right)}} \right)^{\frac{1}{\beta}} \tag{15.10}$$

in which $\Gamma(x) = (x - 1)!$.

15.3 Optimization Procedure of the DA

Like most SI-based optimization algorithms, the DA starts the optimization process by creating a set of random solutions for a given optimization problem. Naturally, the number of initial dragonfly individuals (M) can influence the performance of the DA. The bigger populations increase the chance of finding the global optima, while increasing the calculation time for each iteration, and in turn the entire optimization problem. After determining the positions of dragonflies within the lower and upper boundaries of any given variable, the position of each dragonfly individual is updated in each iteration by calculating the step position vector for each individual dragonfly, using the motions induced by separation, alignment, cohesion, food attraction, and predator distraction. The position updating process is continued iteratively until the termination criterion is satisfied. Table 15.1 lists the characteristics of the DA. Additionally, the flowchart of the DA is shown in Fig. 15.2.

Table 15.1   Characteristics of the DA

General algorithm	Dragonfly algorithm
Decision variable	Dragonfly's position in each dimension
Solution	Dragonfly's position
Old solution	Old position of a dragonfly
New solution	New position of a dragonfly
Best solution	Any dragonfly with the best fitness
Fitness function	Distance from the food source, the predator, center of the swarm, velocity matching, and collision avoidance
Initial solution	Randomly selected position of a dragonfly
Selection	–
Process of generating new solution	Flying with a specific velocity and direction

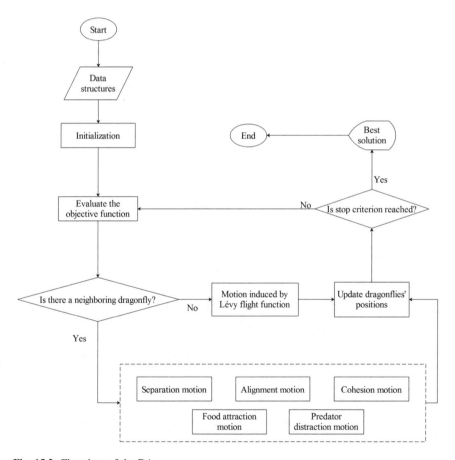

Fig. 15.2   Flowchart of the DA

15.4   Pseudo-Code of the DA

Begin

    Define population size (M)

    Set the iteration counter $t=1$

    Initialize the population by generating X_i for i = 1, 2, 3…, M

    Calculate the objective function values of all dragonflies

    Update the food and the predator's location

    While (the stop criterion is not satisfied)

        For $i=1:M$

            Update neighborhood radii (or update w, s, a, c, f, and e)

            If a dragonfly has at least one neighborhood dragonfly

                Calculate

                    Separation motion

                    Alignment motion

                    Cohesion motion

                    Food attraction motion

                    Predator distraction motion

                    Update position vector

            Else

                Update position vector using the Lévy flight function

            End if

        End for i

        Sort the population/dragonflies from best to worst and find the current best

    End while

    Post-process and visualize the results

End

15.5 Conclusion

This chapter described the dragonfly algorithm (DA) which is a novel, yet newly introduced metaheuristic optimization algorithm. After a brief review of the previous applications of the DA, the standard DA and its mechanism were described. In the final section, a pseudo-code of the standard DA was also presented.

References

Beni, G., & Wang, J. (1993). Swarm intelligence in cellular robotic systems. In: P. Dario, G. Sandini, & P. Aebischer (Eds.), *Robots and biological systems: Towards a new bionics?* Berlin, Heidelberg, New York, NY: Springer.

Bonabeau, E., Dorigo, M., & Theraulaz, G. (1999). *Swarm intelligence: From natural to artificial systems*. New York, NY: Oxford University Press.

Gandomi, A. H., & Alavi, A. H. (2012). Krill herd: A new bio-inspired optimization algorithm. *Communications in Nonlinear Science and Numerical Simulation, 17*(12), 4831–4845.

Gandomi, A. H., Yang, X. S., & Alavi, A. H. (2013). Cuckoo search algorithm: A metaheuristic approach to solve structural optimization problems. *Engineering with Computers, 29*(1), 17–35.

Mirjalili, S. (2016). Dragonfly algorithm: A new meta-heuristic optimization technique for solving single-objective, discrete, and multi-objective problems. *Neural Computing and Applications, 27*(4), 1053–1073.

Reynolds, C. W. (1987). Flocks, herds and schools: A distributed behavioral model. In *Proceedings of the 14th annual conference on computer graphics and interactive techniques*, New York, NY, July 27–31.

Russell, R. W., May, M. L., Soltesz, K. L., & Fitzpatrick, J. W. (1998). Massive swarm migrations of dragonflies (Odonata) in eastern North America. *The American Midland Naturalist, 140*(2), 325–342.

Thorp, J. H., & Rogers, D. C. (Eds.). (2014). *Thorp and Covish's freshwater invertebrates: Ecology and general biology* (Vol. 1). Amsterdam, Netherland: Elsevier.

Wikelski, M., Moskowitz, D., Adelman, J. S., Cochran, J., Wilcove, D. S., & May, M. L. (2006). Simple rules guide dragonfly migration. *Biology Letters, 2*(3), 325–329.

Yang, X. S. (2010). *Nature-inspired metaheuristic algorithms*. Frome, UK: Luniver Press.

Printed in the United States
By Bookmasters